Neural Computing for Optimization and Combinatorics

NEURAL COMPUTING FOR OPTIMIZATION AND COMBINATORICS

Editors

Yoshiyasu Takefuji
Case Western University, USA
Keio University, Japan

Jun Wang
University of North Dakota, USA

World Scientific
Singapore • New Jersey • London • Hong Kong

Published by

World Scientific Publishing Co. Pte. Ltd.

P O Box 128, Farrer Road, Singapore 912805

USA office: Suite 1B, 1060 Main Street, River Edge, NJ 07661

UK office: 57 Shelton Street, Covent Garden, London WC2H 9HE

NEURAL COMPUTING FOR OPTIMIZATION AND COMBINATORICS

ISBN 981-02-1314-X

This book is printed on acid-free paper.

Printed in Singapore by Uto-Print

PREFACE

This book is intended for undergraduate or graduate students, or applied mathematicians/engineers who would like to study neural computing for optimization and combinatorics. This book is composed of twelve chapters including N-queen in Chapter 1, crossbar scheduling in Chapter 1, gate packing in Chapter 2, maximum clique in Chapters 3 and 4, multilayer channel routing in Chapter 5, job-shop scheduling in Chapter 6, BIBD in Chapter 7, RNA structure prediciton in Chapter 8, missionaries and cannibals in Chapter 9, FLN learning for optimization in Chapter 10, identification and control in Chapter 11, and Ramsey numbers in Chapter 12. Chapter 1 is based on the paper published in Proceedings of World Congress on Neural Networks in 1993 and also on the paper published in Biological Cybernetics, 64, pp. 353–356, 1991. Chapters 2 and 3 are based on the Case Western Reserve University dissertation published in 1991. Chapter 4 is based on the paper published in Journal of Parallel Distributed Computing, 14, pp. 340–344, 1992. Chapter 5 is based on the paper published in Proceedings of the International Conference on Automation, Robotics, and Computer Vision, Sept. 1992. Chapter 6 is based on the paper published in Proceedings of the International Conference on Networks in 1988. Chapter 7 is based on the paper published in IEEE Transactions on Circuits and Systems, 39, No. 4, pp. 243–249, 1992. Chapter 8 is based on the paper published in BioSystems, 27, No. 2, pp. 85–96, 1992. Chapter 9 is a new paper. Chapter 10 is based on the paper published in IEEE Computer, 25, No. 5, pp. 76–79, 1992. Chapter 11 is based on the paper published in International Journal of Analog Integrated Circuits and Signal Processing, 2, No. 4, pp. 353–363, 1992. Chapter 12 is based on the paper published in International Journal of Analog Integrated Circuits and Signal Processing, 2, No. 4, pp. 313–322, 1992.

Yoshiyasu Takefuji
Jun Wang

ACKNOWLEDGMENTS

We would like to thank many friends who have contributed to this book including Professor Yoh-Han Pao at Case Western Reserve University, Professor Arieh Zaritsky at Ben-Gurion University of the Negev, Professor Kurokawa at The National Defense Academy, Professor Phillips at Case Western Reserve University, and my former students including Professor Simon Foo, Dr. Kuo Chun Lee, and Dr. Nobuo Funabiki. I also would like to thank Mr. Kazuhiro Tsuchiya, Mr. Kyotaro Suzuki, Mr. Akira Hanada, and Mr. Yoshiaki Araki. I also thank Michiyo Takefuji for her patience who allows me to finish this book.

CONTENTS

Chapter 1

N-QUEEN AND CROSSBAR PROBLEMS

YOSHIYASU TAKEFUJI
Dept. of Electrical Engineering and Applied Physics
Case Western Reserve University
Cleveland, OH 44106, USA
and
Faculty of Environmental Information
Keio University, Fujisawa 252, Japan

A neural network approach and a backtracking approach to N-queen problems were examined. The goal of an N-queen problem is to locate N queens on an $N \times N$ chessboard where no two queens attack to each other. Two programs based on the artificial neural network algorithm and the backtracking algorithm were developed on a Macintosh machine and their performance was empirically tested. The crossbar switch scheduling problem (N-rook problem) similar to the N-queen problem is also discussed to illustrate another example of neural network computing for the real-world application.

1.1 N-QUEEN PROBLEMS

The 8-queen problem was proposed in 1848 and it was investigated by several famous mathematicians including C. F. Gauss in 1850 where the goal of the problem is to place 8 queens on an 8×8 chessboard in mutually nonattacking positions. The 8-queen problem has been used as a benchmark problem so as to demonstrate divide-and-conquer methods (Abramson and Yung 1989), trial-and-error methods including back-tracking algorithms (Bitner and Reingold 1975) (Stone and Stone 1987), and other methods (Yaglom and Yaglom 1964) (Kale 1990). Although few parallel algorithms have been proposed (Filman and Friedman 1984) (Finkel and Manber 1987) (Abramson and Yung 1989), there exists no satisfactory parallel algorithm. Page et al. presented the first deterministic parallel algorithm based on the Hopfield neural network (Page and Tagliarini 1987). Akiyama et al. proposed an $O(N^2)$ parallel algorithm based on the stochastic neural network model (Akiyama et al. 1989). However the state of the system based on the existing neural network algorithms may converge to the unacceptable local minimum. In other words, their solution quality dramatically degrades with the problem size.

The goal of a general N-queen problem is to locate N queens on an $N \times N$ chessboard where any pair of queens does not command each other. Note that a queen commands vertically, horizontally, and diagonally as shown in Fig. 1.1.

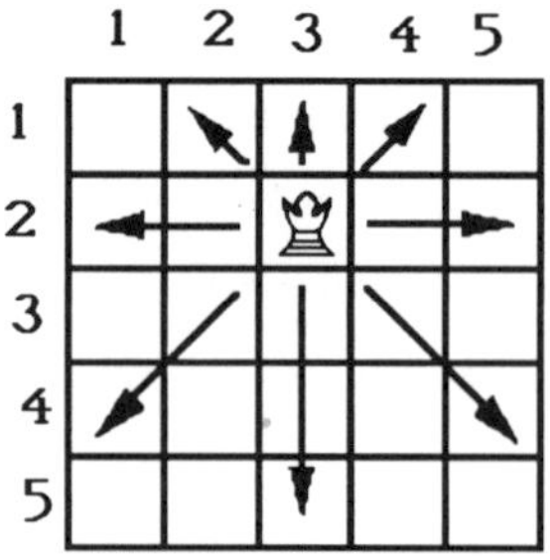

Fig. 1.1 How a queen commands.

1.2 MATHEMATICAL NEURAL NETWORK MODELS

The mathematical model of the artificial neural network consists of two components; neurons and synaptic links. The output signal transmitted from a neuron propagates to other neurons through the synaptic links. The state of the input signal of a neuron is determined by the linear sum of weighted input signals from the other neurons where the respective weight is the strength of the synaptic links. Every artificial neuron has the input U and the output V. The output of the *i*th neuron is given by $V_i=f(U_i)$ where f is called the neuron's input/output function. In this chapter we introduce a hysteresis McCulloch-Pitts function as the neuron's input/output function.

The interconnections between the *i*th neuron and other neurons are determined by the motion equation. The change of the input state of the *i*th neuron is given by the partial derivatives of the computational energy function E with respect to the output of the *i*th neuron where E follows an n-variable function: $E(V_1, V_2,\ldots,V_n)$. The motion equation of the *i*th neuron is given by:

$$\frac{dU_i}{dt} = - \frac{\partial E(V_1, V_2, \ldots, V_n)}{\partial V_i} \tag{1}$$

The goal of the artificial neural network for solving optimization problems is to minimize the fabricated computational energy function E in Eq. (1). The energy function not only determines how many neurons should be used in the system but also the strength of the synaptic links between neurons. It is constructed by considering the necessary and sufficient constraints and/or the cost function from the given problem. It is usually easier to build the motion equation than the energy function. From Eq. (1) the energy function E can be obtained:

$$E = \int dE = -\int \frac{dU_i}{dt} dV_i \qquad (2)$$

The artificial neural network provides a parallel gradient descent method to minimize the fabricated energy function E. Convergence theorems/proofs of the artificial neural network are given in (Takefuji 1992). The state of the McCulloch-Pitts neural network is allowed to converge to the local minimum where the convergence speed is relatively faster than that of the sigmoid neural network. However the McCulloch-Pitts neural network sometimes generates undesirable oscillatory behaviors. In order to suppress oscillatory behaviors, the hysteresis McCulloch-Pitts neuron model as shown in Fig. 1.2 is introduced where the input/output function of the *i*th hysteresis McCulloch-Pitts neuron is given by:

$V_i=1$ if $U_i>$UTP (Upper Trip Point)
$=0$ if $U_i<$LTP (Lower Trip Point)
unchanged otherwise (3)

where UTP is always larger than LTP. Because of suppressing the oscillatory behavior it shortens the convergence time consequently.

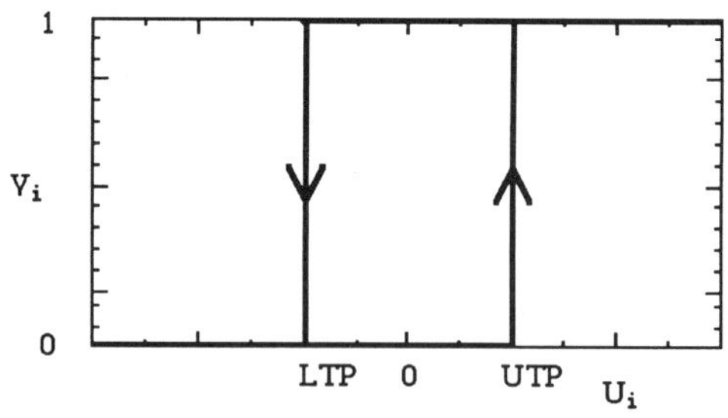

Fig. 1.2 Hysteresis McCulloch-Pitts input/output function.

1.3 N-QUEEN NEURAL NETWORK

It is well understood that one and only one queen must be located per row and per column so as to locate N queens on an $N \times N$ chessboard. In other words more than one queen should not be located per row and per column. The condition of the constraints or the violations gives the interconnections of the artificial neural network. An $N \times N$ neural array is prepared for the N-queen problem where each element of the neural array represents a square of the $N \times N$ chessboard. In other words, N^2 neurons are used for the N-queen problem. The output state of the *ij*th neuron gives the location of a queen in the *i*th row and the *j*th column. In other words, $V_{ij}=1$ means that a queen is located in the *i*th row and the *j*th column. $V_{ij}=0$ means that no queen is located in the *i*th row and the *j*th column. For example, to locate a queen in the second row and the third column as shown in Fig. 1, V_{23} must be $V_{23}=1$. The motion equation of the *ij*th neuron for the N-queen problem is given by:

$$\frac{dU_{ij}}{dt} = -A\left(\sum_{k=1}^{N} V_{ik} - 1\right) - A\left(\sum_{k=1}^{N} V_{kj} - 1\right) - B \sum_{\substack{1 \le i-k, j-k \le N \\ k \ne 0}} V_{i-k,j-k} - B \sum_{\substack{1 \le i-k, j+k \le N \\ k \ne 0}} V_{i-k,j+k} \qquad \text{for } i,j=1,..,5 \tag{4}$$

In Eq. (4) the first term gives the row constraint such that one and only one queen/neuron should be located/fired in the *i*th row. The second term describes the column constraint such that one and only one queen/neuron should be located/fired in the *j*th column. The third and the fourth term represent the diagonal constraints such that no pair of queens should not diagonally command each other. The following two terms must be added to Eq. (4) to enable the state of the system to escape from the local minimum and to converge to the global minimum or the acceptable solution:

$$+Ch\left(\sum_{k=1}^{N} V_{ik}\right) +Ch\left(\sum_{k=1}^{N} V_{kj}\right) \tag{5}$$

where h(x) is 1 if x=0, and 0 otherwise. The function of h(x) is called hill-climbing term. In the local minimum no neurons/queens are fired/located on the *i*th row or the *j*th column.

1.4 N-QUEEN SIMULATION PROGRAMS

In order to numerically solve the partial differential equation or the differential equation, the first order Euler method is widely used where it is the simplest among the existing numerical methods. Based on the first order Euler method, the value of U(t+1) is determined by U(t) and ΔU(t):

$$U(t+1)=U(t)+\Delta U(t)\cdot \Delta t \tag{6}$$

where ΔU(t) is given by Eq. (4). With $\Delta t=1$, Eq. (6) for N-queen problems (where a two dimensional neural array is used) is given by:

$$U_{ij}(t+1)=U_{ij}(t)+\Delta U_{ij}(t) \quad \text{for } i,j=1,\ldots,N \tag{7}$$

In order to solve N-queen problems, the initial values of $U_{ij}(0)$ for i,j=1,...,N are uniform-randomly given and Eq. (7) is iteratively used until the state of the system reaches the equilibrium state. However the equilibrium state is not always equivalent to the acceptable solution. In the N-queen problem, the condition of the program termination is given by:

$$\Delta U_{ij}(t)=0 \quad \text{for } i,j=1,\ldots,N \tag{8}$$

The condition of Eq. (8) implies that the constraints are all satisfied. In other words, no violations can be found on the condition. Eq. (7) is either sequentially updated or parallelly updated. In parallel computing there are a synchronous parallel model and an asynchronous parallel model. In the synchronous parallel model, clock signals or handshaking signals are used for synchronous communications between processing elements (neurons). In the asynchronous parallel model the state of every processing element (neuron) is updated asynchronously without using any clock/handshaking signals.

Fig. 1.3 shows a Turbo Pascal program on any Macintosh machine. The program is to simulate the sequential N-queen neural network model. Fig. 1.4 shows a Turbo Pascal program for simulating the synchronous parallel N-queen neural network model where the hysteresis McCulloch-Pitts neuron is used. UTP=3 and LTP=0 are used for hysteresis. It takes about 5.5 minutes to solve a 80-queen problem on a Macintosh SE/30 where it only requires 28 iteration steps. If 6,400 processing elements are provided and assume that each processing element has the processing speed of one million state-updates per second (1 MSUPS), it will take 28 micro seconds to solve the 80-queen problem. Fig. 1.5 shows one of the solutions for the 80-queen problem.

```
program sequential_queen;
uses memtypes, quickdraw,osintf,ToolIntf;
type twod_int= array[1..89,1..89] of integer;
var
A,B,C,t,h,i,j,k,sum_column,sum_row,diagonal1,diagonal2,conf,diag,seed,row,length
,max:integer;
U,V:twod_int; tempRect : Rect;
begin
  A:=1; B:=1; C:=1;
  writeln('Please define the queen problem size (6-89).'); readln(max);
  writeln('Please input a seed(0-9999).'); readln(seed);
  for i:=1 to seed do U[1,1]:=random; length:=trunc(270/max);
  for i:=1 to max do
    for j:=1 to max do
    begin U[i,j]:=-trunc(abs(random/4000)); if U[i,j]>0 then V[i,j]:=1 else V[i,j]:=0;
end;
  t:=0; diag:=1;  {******** Main program ********}
  while ((not keypressed) and (diag>0) and (t<2000)) do
```

```
    begin diag:=0;
      for i:=1 to max do
       for j:=1 to max do
        begin
          sum_column:=0;sum_row:=0;
          for k:=1 to max do begin sum_row:=sum_row+V[i,k];
          sum_column:=sum_column+V[k,j]; end;
          diagonal1:=0; k:=1;
         while((j+k)<=max) and ((i-k)>=1)
         do begin diagonal1:=diagonal1+V[i-k,j+k]; k:=k+1; end;
          k:=1;
         while((j-k)>=1) and ((i+k)<=max)
         do begin diagonal1:=diagonal1+V[i+k,j-k]; k:=k+1; end;
          diagonal2:=0; k:=1;
         while((j+k)<=max) and ((i+k)<=max)
         do begin diagonal2:=diagonal2+V[i+k,j+k]; k:=k+1; end;
          k:=1;
        while((j-k)>=1) and ((i-k)>=1) do begin diagonal2:=diagonal2+V[i-k,j-k];
          k:=k+1; end;
        h:=0;conf:=1;
        if (sum_column=0) then h:=1; if (sum_row=0) then h:=h+1;
      if((sum_column+sum_row=2) and (diagonal1<2) and (diagonal2<2)) then
       conf:=0;
      U[i,j]:=U[i,j]-A*(sum_row+sum_column-2)-B*(diagonal1+diagonal2)+C*h;
       if (U[i,j]>15) then U[i,j]:=15; if (U[i,j]<-5) then U[i,j]:=-5;
           if U[i,j]>0 then V[i,j]:=1 else V[i,j]:=0;
      diag:=diag+conf;
    end;{******** The end of i and j loop ********}
     t:=t+1; if (t mod 20 <5) then C:=4 else C:=2; write(t);
 end; {******** The end of while loop ********}
clearscreen;
for i:=1 to max do
 for j:=1 to max do
  if (V[i,j]=1) then
    begin
     SetRect(tempRect,0,0,length,length); OffsetRect(tempRect,(i-1)*length,(j-
1)*length); PaintRect(tempRect);
    end;
  MoveTo(0,0); LineTo(0,max*length); MoveTo(0,max*length);
LineTo(max*length,max*length);
  MoveTo(max*length,max*length); LineTo(max*length,0);
MoveTo(max*length,0); LineTo(0,0);
 for i:=1 to max do begin
    MoveTo(i*length,0); LineTo(i*length,max*length); MoveTo(0,i*length);
```

```
LineTo(max*length,i*length); end;
gotoXY(50,1);writeln('No. of steps=',t);readln;readln;
end. {******** The end of program ********}
```

Fig. 1.3 A program of the sequential N-queen neural network.

```
program sync_parallel_queen;
uses memtypes, quickdraw,osintf,ToolIntf;
type twod_int= array[1..89,1..89] of integer;
var
A,B,C,t,h,i,j,k,sum_column,sum_row,diagonal1,diagonal2,conf,diag,seed,row,length
,max:integer;
U,V:twod_int; tempRect : Rect;
begin
   A:=1; B:=1; C:=1;
   writeln('Please define the queen problem size (6-89).'); readln(max);
   writeln('Please input a seed(0-9999).'); readln(seed);
   for i:=1 to seed do U[1,1]:=random; length:=trunc(270/max);
   for i:=1 to max do
     for j:=1 to max do
     begin U[i,j]:=-trunc(abs(random/1600)); if U[i,j]>0 then V[i,j]:=1 else V[i,j]:=0;
end;
   t:=0; diag:=1;   {******** Main program ********}
   while ((not keypressed) and (diag>0) and (t<2000)) do
     begin diag:=0;
       for i:=1 to max do
        for j:=1 to max do
         begin
           sum_column:=0;sum_row:=0;
          for k:=1 to max do begin sum_row:=sum_row+V[i,k];
sum_column:=sum_column+V[k,j]; end;
          diagonal1:=0; k:=1;
         while((j+k)<=max) and ((i-k)>=1) do begin diagonal1:=diagonal1+V[i-
k,j+k]; k:=k+1; end;
          k:=1;
         while((j-k)>=1) and ((i+k)<=max) do begin diagonal1:=diagonal1+V[i+k,j-
k]; k:=k+1; end;
          diagonal2:=0; k:=1;
         while((j+k)<=max) and ((i+k)<=max) do begin
diagonal2:=diagonal2+V[i+k,j+k]; k:=k+1; end;
```

```
          k:=1;
        while((j-k)>=1) and ((i-k)>=1) do begin diagonal2:=diagonal2+V[i-k,j-k];
k:=k+1; end;
       h:=0;conf:=1;
       if (sum_column=0) then h:=1; if (sum_row=0) then h:=h+1;
      if((sum_column+sum_row=2) and (diagonal1<2) and (diagonal2<2)) then
conf:=0;
      U[i,j]:=U[i,j]-A*(sum_row+sum_column-2)-B*(diagonal1+diagonal2)+C*h;;
       if (U[i,j]>15) then U[i,j]:=15; if (U[i,j]<-20) then U[i,j]:=-20;
       diag:=diag+conf;
     end;{******** The end of i and j loop ********}
for i:=1 to max do
     for j:=1 to max do
     begin if U[i,j]>3 then V[i,j]:=1; if U[i,j]<0 then V[i,j]:=0; end;
      t:=t+1; if (t mod 20 <5) then C:=4 else C:=2; write(t);
 end; {******** The end of while loop ********}
 clearscreen;
 for i:=1 to max do
  for j:=1 to max do
   if (V[i,j]=1) then
     begin
      SetRect(tempRect,0,0,length,length); OffsetRect(tempRect,(i-1)*length,(j-
1)*length); PaintRect(tempRect); end;
  MoveTo(0,0); LineTo(0,max*length); MoveTo(0,max*length);
LineTo(max*length,max*length);
  MoveTo(max*length,max*length); LineTo(max*length,0);
MoveTo(max*length,0); LineTo(0,0);
 for i:=1 to max do
   begin
    MoveTo(i*length,0); LineTo(i*length,max*length); MoveTo(0,i*length);
LineTo(max*length,i*length); end;
gotoXY(50,1);writeln('No. of steps=',t);readln;readln;
end. {******** The end of program ********}
```

Fig. 1.4 A program of the synchronous-parallel N-queen neural network.

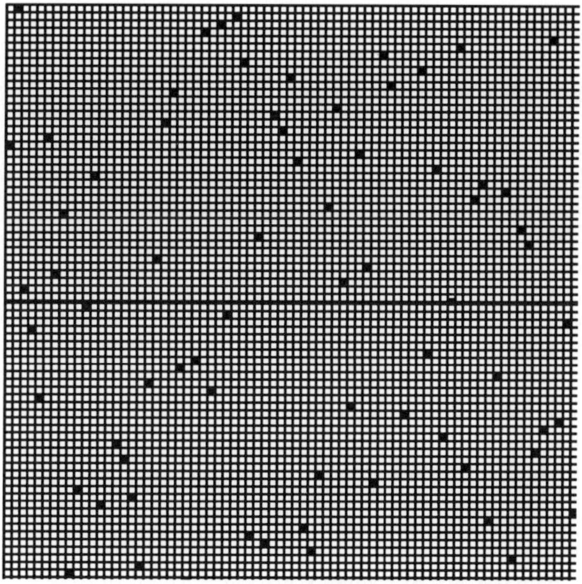

Fig. 1.5 One of the solutions for the 80-queen problem.

1.5 BACKTRACKING ALGORITHM USING PROLOG

The N-queen program based on the backtracking algorithm was written in Prolog. Fig. 1.6 shows the Prolog program for N-queen problems. To obtain the whole solutions for the 8-queen problem, type "q(8)." To obtain a single solution of the 8-queen problem, type "queen(8,X)." and type ";" to get another solution.

```
q(N) :-
    asserta(qc(0)),
    queen(N,L),
    call(qc(X)),
    abolish(qc,1),
    X1 is X+1,
    asserta(qc(X1)),
    write('No. '),write(X1),nl,
    disp(L,N),
    nl,
```

```
        fail.
q(_) :- retract(qc(N)), write('Total = '), write(N),nl.
queens(N) :- queen(N,L), write(L),nl,fail.
queen(N,L) :-  generate(N,L1), queen1(N,L1,[],L,[],[]).
generate(0,[]).
generate(N,[N|L]) :- N>0, N1 is N-1, generate(N1,L).
queen1(_,[],L,L,_,_) :- !.
queen1(M,S,L1,L,C,D) :-
        select(S,A,S1),
        C1 is M+A,
        notmem(C1,C),
        D1 is M-A,
        notmem(D1,D),
        M1 is M-1,
        queen1(M1,S1,[A|L1],L,[C1|C],[D1|D]).
select([A|L],A,L).
select([A|L],X,[A|L1]) :- select(L,X,L1).
notmem(_,[]) :- !.
notmem(A,[B|L]) :- A=\=B, notmem(A,L).
disp([],_) :- !,nl.
disp([A|X],N) :- disp1(A,N), nl,!, disp(X,N).
disp1(1,X) :- !, write(' Q'), disp2(X).
disp1(X,Y) :- X1 is X-1, Y1 is Y-1, write(' .'), disp1(X1,Y1).
disp2(1) :- !.
disp2(X) :- X1 is X-1, write(' .'), disp2(X1).
```

Fig. 1.6 Prolog program for N-queen problems.

The computation time of the program in Fig. 1.6 exponentially increases with the problem size. For example, the 12-queen problem and the 15-queen problem required 5 seconds and 90 seconds respectively to obtain a single solution on Macintosh SE/30. It took 150 minutes to obtain a solution of the 20-queen problem while the neural network program required less than 10 seconds on Macintosh SE/30. Generally the neural network program running on a sequential machine requires $O(N^2)$ time for solving the N-queen problem and nearly O(1) time on a parallel machine using N processing elements based on our simulation result.

1.6 CROSSBAR SWITCH SCHEDULING

In time-multiplex communication systems, crossbar packet switches route traffic from the input to output where a message packet is transmitted from the source to the destination. The randomly incoming traffic must be controlled and scheduled to eliminate conflict at the crossbar switch where the conflict is that two or more users may simultaneously access to a single output. The goal of the traffic-scheduling for time-multiplex crossbar switches is not only to maximize the throughput of packets through a crossbar switch but also to minimize packet blocking probability and packet waiting time.

A request for packet transmission through an n $\times$ n crossbar is described by an n $\times$ n traffic matrix T. In the traffic matrix T, each element t_{ij} represents a request of packets from input i to output j. For example, $t_{ij}=0$ means that there is no packet to be transmitted on the jth output line from the ith input line. $t_{ij}=1$ means that at least one packet on the *i*th input line should be transmitted on the *j*th output line of the crossbar. Fig. 1.7(a) shows a 5 $\times$ 5 crossbar switch where 7 packets at least are requested: $t_{11}=t_{22}=t_{24}=t_{35}=t_{42}=t_{45}=t_{53}=1$. 5 packets are maximally transmitted simultaneously through the 5 $\times$ 5 crossbar switch. Fig. 1.7(b) depicts the solution of the problem in Fig. 1.7(a) where two requests t_{22} and t_{45} are rejected.

Fig. 1.7 (a) 5 $\times$ 5 crossbar switch problem

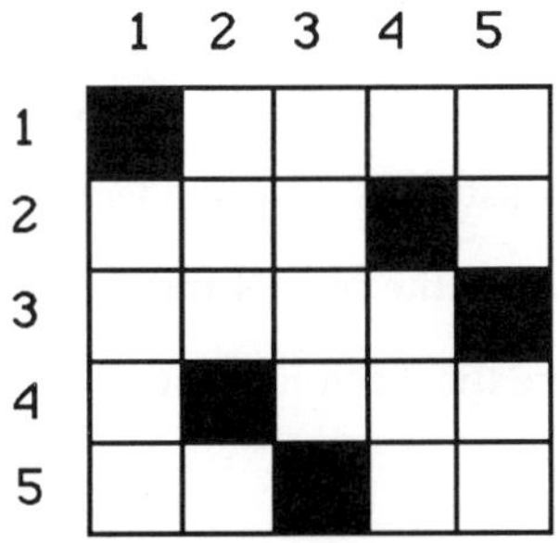

Fig. 1.7 (b) One of the solutions.

In 1979 Inukai at COMSAT lab. proposed the $O(N^2)$ sequential algorithm for the $N \times N$ crossbar switch problem (Inukai 1979). In 1989 Rose at AT&T Bell lab. presented the O(N) parallel algorithm based on a cellular automaton where N^2 processing elements are used for solving an $N \times N$ traffic matrix problem (Rose 1989). Chen, Mavor, Denyer, and Renshaw proposed the $O(N^2)$ sequential algorithm of traffic routing problems for the multiprocessor system (Chen et al. 1990). They proved that the problem is NP-complete (Chen et al. 1990). In 1989 Marrakchi and Troudet at Bellcore proposed the $N \times N$ neural network algorithm based on Hopfield network model (Marrakchi and Troudet 1989). However with the Hopfield neural network, the state of the system is forced to converge to the local minimum. In other words, the solution quality drastically degrades with the problem size. The system uses an $N \times N$ neural network array for solving an $N \times N$ crossbar switch problem where the output of the ijth hysteresis McCulloch-Pitts neuron V_{ij}.

The proposed hysteresis McCulloch-Pitts binary neural network not only maximizes the throughput of packets through a crossbar switch but also minimizes packet blocking probability and packet waiting time. The constraints on an $N \times N$ crossbar switch are that no two inputs may be connected to the same output simultaneously and that no one input may be connected to more than one output simultaneously. In other words, no two packets should share the same row and the column of the traffic matrix. The system simulator was developed based on the proposed model. A large number of simulation results are shown and demonstrated in order to support the effects of hysteresis.

The constraints are considered that no two packets should share the same row and the column of the $N \times N$ traffic matrix. Our system uses an $N \times N$ neural network array where the motion equation of the *ij*th neuron is given by:

$$\frac{dU_{ij}}{dt} = -A\left(\sum_{k=1}^{N} V_{ik} - 1\right) - A\left(\sum_{k=1}^{N} V_{kj} - 1\right) + Bh\left(\sum_{k=1}^{N} V_{ik}\right) + Bh\left(\sum_{k=1}^{N} V_{kj}\right) \quad (9)$$

where h(x) is 1 if x=0, 0 otherwise. Note that coefficients A and B are constant integers. The first term and the second term are the row constraint and the column constraint respectively. The first term forces one and only one neuron to be fired per row. If no neuron is fired per row or per column then it will perform excitatory forces. If more than one neuron are fired per row or per column then it will act inhibitory forces. The third term and the last term are the row hill-climbing term and the column hill-climbing term respectively. The hill-climbing terms are activated only when local conflicts (where no neuron is fired per row or column) are detected. If there is no conflict the hill-climbing terms will perform no operation. In other words, the local conflicts are resolved by the hill-climbing terms where the *ij*th neuron is encouraged to fire if the neuron has conflicts.

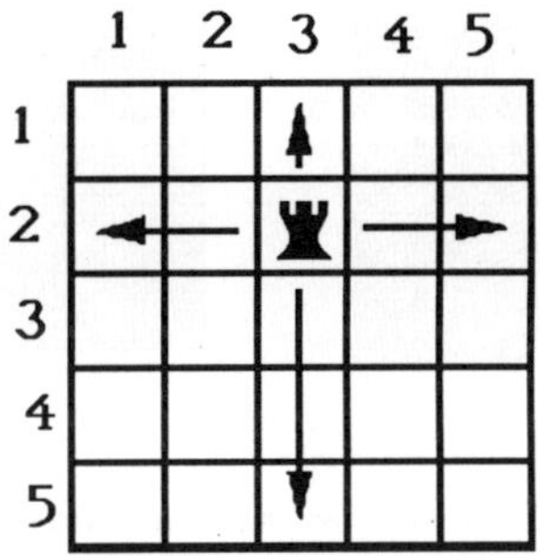

Fig. 1.8 How a rook commands.

The reader might associate with the N-queen problem from the crossbar switch problem since they are very similar to each other except the diagonal constraints in N-queen problems. Eq. (4) with the hill-climbing term in Eq. (5) is exactly equivalent to Eq. (9) after removing B-term (diagonal terms) from Eq. (9). In other words, the crossbar switch scheduling problem is considered as N-rook problem where a rook commands horizontally and vertically. The positions of the rooks are equivalent to the state of the traffic matrix. The goal of the $N \times N$ crossbar switch scheduling problem is to place N mutually nonattacking rooks on an $N \times N$

chessboard. The maximum number of rooks on the N × N chessboard is equivalent to N^2. Fig. 1.8 shows how a rook commands.

Fig. 1.9(a), and 1.9(b) show the 70 × 70 crossbar problem and one of the solutions respectively. The following parameters were used in our simulation: A=1, B=4 if (t mod 10)<2, 1 otherwise. On Macintosh SE/30 it took 55 seconds for our simulator to obtain the solution shown in Fig. 1.9. The intensive study on crossbar switch scheduling was depicted in (Takefuji and Lee 1991).

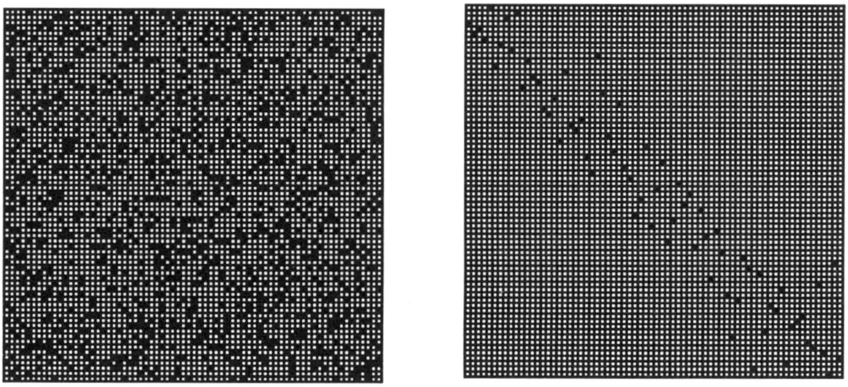

(a) The problem (b) one of the solutions

Fig. 1.9 The 70 × 70 crossbar switch problem and one of the solutions.

1.7 REFERENCES

Abramson B., and Yung M., (1989), Divide and conquer under global constraints: a solution to the N-queens problem, J. of Parallel and Distributed Computing, 6, 649-662.

Akiyama Y., Yamashita A., Kajiura M., and Aiso H., (1989), Combinatorial optimization with Gaussian machines. Proc. of the Int. Joint Conf. on Neural Networks.

Bitner J, and Reingold E. M., (1975), Backtrack programming techniques, Comm.

ACM 18, 651-655.

Filman R. E., and Friedman D. P., (1984), Coordinated computing: tools and techniques for distributed software, (McGraw-Hill, New York).

Finkel R., and Manber U., (1987), DIB-a distributed implementation of backtracking, ACM Trans. Programming Language Systems, 9, 235-256.

Funabiki, N. and Takefuji, Y., (1993), Comparisons of seven neural network models on traffic control problems in multistage interconnection networks, IEEE Trans. on Computers, 42, 4, 497-501.

Hebb D. O, (1949), The organization of behavior (New York: Wiley).

Hopfield J. J., and Tank D. W., (1985), Neural Computation of Decisions in Optimization Problems. Biological Cybernetics, 52, 141-152.

Kale L.V., (1990), An almost perfect heuristic for the N nonattacking queens problem, Information Processing Letters, 34, 173-178.

McCulloch W. S., and Pitts W. H., (1943), A logical calculus of ideas immanent in nervous activity. Bulletin of Mathematical Biophysics, 5, 115.

Minsky M, and Papert S., (1969), Perceptrons (MIT Press).

Page E., and Tagliarini G. A, (1987), Solving constraint satisfaction problems with neural networks, Proc. of IEEE first Int. Conf. on Neural Networks.

Paielli R. A., (1988), Simulation tests of the optimization method of Hopfield and Tank using neural networks, NASA Technical Memorandom 101047.

Rosenblatt F., (1962), Principles of neurodynamics (New York: Spartan).

Stone H., and Stone J. M., (1987), Efficient search techniques-an empirical study of N-queens problem, IBM J. Res. Develop., 31, 4.

Takefuji Y., and Lee K. C., (1991), Artificial neural networks for four-coloring problems and k-colorability problems, IEEE Trans. on Circuits and Systems, 38, 3, pp.326-333.

Takefuji Y. and Lee K. C, (1990), A super parallel sorting algorithm based on neural networks, IEEE Trans. on Circuits and Systems, 37, 11, 1425-1429.

Widrow B, and Hoff M. E., (Sept. 1960), Adaptive switching circuits. Proc. of IREWESCON Convention Record.

Wilson G. V. and Pawley G. S., (1988), On stability of the Travelling Salesman Problem Algorithm of Hopfield and Tank. Bio. Cybern., 58, 63-70.

Yaglom A. M., and Yaglom I. M., (1964), Challenging mathematical problems with elementary solutions, (Holden-Day, San Francisco).

Chen W., Mavor J., Denyer P. B., Renshaw D. (1990) Traffic routing algorithm for serial superchip system customisation. IEE Proc., 137, Pt. E, 1

Hoffman G. W., Benson M. W. (1986) Neurons with hysteresis form a network that can learn without any changes in synaptic connection strengths. Proc. of AIP Conf. on Neural Networks for Computing (Ed. J. S. Denker, AIP 1986)

Hopfield J. J. (1982) Neural networks and physical systems with emergent collective computational abilities. Proc. Natl. Acad. Sci. USA, 79, 2554-2558

Hopfield J. J., Tank D. (1986) Computing with neural circuits: A model. Science, 233, 625-633

Inukai T. (1979) An efficient SS/TDMA time slot assignment algorithm. IEEE Trans. on Commun., 27, 1449-1455

Keeler J. D., Pichler E. E., Ross J. (1989) Noise in neural networks: thresholds, hysteresis, and neuromodulation of signal-to-noise. Proc. Natl. Acad. Sci. USA, 86, 1712-1716

Marrakchi A., Troudet T. (1989) A Neural Net Arbitrator for Large Crossbar Packet-Switches. IEEE Trans. on Circuits and Systems, 36, 7

McCulloch W. S., Pitts W. (1943) A logical calculus of the ideas imminent in nervous activity. Bull. Math. Biophys., 5, 115-133

Rose C. (1989) Rapid Optimal Scheduling for Time-Multiplex Switches Using a Cellular Automaton. IEEE Trans. on Commun., 37, 500-509

Segundo J. P., Martinez O. D. (1985) Dynamic and Static hysteresis in Crayfish stretch receptors. Biological Cybernetics, 52, 291-296

Simic B. G. (1986) SPIE, 634, 195

Takefuji Y., Lee K. C. (1989) A near-optimum parallel planarization algorithm. Science, 245, 1221-1223

Takefuji Y., Lee K. C. (1990a) A parallel algorithm for tiling problems. IEEE

Trans. on Neural Networks, 1, 1, 143-145

Takefuji Y., Lee K. C. (1990b) A super parallel sorting algorithm based on neural networks. IEEE Trans. on Circuits and Systems, 37, 11

Takefuji Y., Lee K. C. (1991) An hysteresis binary neuron: a model suppressing the oscillatory behaviors of neural dynamics. Biological Cybernetics, 64, 353-356

Chapter 2

GATE PACKING PROBLEMS

FUYAU LIN
Santa Clara University, Computer Engineering, Santa Clara, CA 95053, USA

KUO-CHUN LEE
Cirrus Logic, Inc., R & D Department, Fremont, CA 94538, USA

YOSHIYASU TAKEFUJI
Dept. of Electrical Engineering and Applied Physics
Case Western Reserve University
Cleveland, OH 44106, USA
and
Faculty of Environmental Information
Keio University, Fujisawa 252, Japan

A parallel algorithm for gate packing problems is presented in this chapter. The layout problem of gate matrices and one-dimensional logic arrays is composed of two major tasks: one to find a permutation of gates which minimizes the number of tracks required and the other to layout/pack gates based on the ordering. The algorithm presented here can pack n gates within $O(1)$ time, while the conventional near-optimum algorithm needs $O(n^2)$ time. The simulation results show that the increase of the problem size does not degrade the solution quality.

2.1 INTRODUCTION

Digital design is an exploration of the space of possible choices and a justification of time constraints. VLSI (Very Large Scale Integration) is concerned with the means and methods for packing a number of electronic components onto a small chip. The layout design problem of a large scale integration circuit was introduced by Weinberger in 1967[6]. A minimum track placement problem is analogous to the general layout compaction problem. The problem is known as NP-complete[3]. The layout problem can be divided into two tasks. The first task is to find a permutation of the gates or the columns which minimizes the number of tracks to pack the gates. Several algorithms for ordering the gates have been reported by [1,2,7,8]. The second task is to pack and layout gates to minimize the area of the chip. Current large-scale design problems are time-consuming so that they cannot be solved promptly by the conventional computer. There is a demand to employ the power of multiple processors working in parallel. To exploit parallelism with the VLSI technology we must develop new algorithms. Hashimoto and Stevens[4] has proposed the left-edge first algorithm in 1971 which needs $O(n^2)$ time. No parallel algorithm for packing gates has been reported since then.

In this chapter a parallel algorithm is presented which can layout and pack n gates within $O(1)$ time. The algorithm uses a large number of processing elements. The proposed algorithm is the implementation of the modified Hopfield's computational artificial neural network. The neural network consists of m clusters where each cluster is composed of n processing elements (PE). The proposed algorithm uses a two dimensional $m \times n$ processor network where m and n are the number of tracks and the number of gates respectively. The output of the xith

PE in the two dimensional processor network follows the winner-take-all binary function:

$$V_{xi} = 1 \text{ if } U_{xi} > 0 \text{ and } U_{xi} = \max(U_{ki}) \text{ for } k = 1, \ldots, m \quad (1)$$
$$= 0 \text{ otherwise}$$

where U_{xi} and V_{xi} are the input and output of the processing element *xi*, respectively. All the PE are running concurrently to calculate their *U*s and *V*s.

2.2 PRELIMINARY

In a VLSI circuit, a complex logic function can be realized with gates interconnected as a one-dimensional array. In the detailed layout, in-gate wiring is performed along the vertical columns and between-gates connections are made horizontally. The width of each gate and the separation distance between two neighboring gates are *ad hoc* with respect to physical constraints. The total horizontal dimension of an array is considered fixed. The vertical tracks determine the total chip area which must be minimized. The problem of finding a minimum track placement can be analogous to the general layout compaction problem. Compaction in the vertical direction can be regarded as a column sequencing problem and compaction in the horizontal direction as a net assignment problem. Given a net sets $X(n_i)$, $n_i \in N$ of a circuit, find a pair of gates and net assignment functions $f: G \rightarrow C$ and $h : N \rightarrow R$, respectively, such that the cardinality of R is minimum. For example, a net list $t_1 = \{1, 2, 3\}$, $t_2 = \{3,5\}$, $t_3 = \{4,7\}$, $t_4 = \{2,4,5\}$, $t_5 = \{6,7,8\}$ will require 5 tracks without compaction, but we can easily identify a better placement using only 3 tracks as shown in Fig. 1.

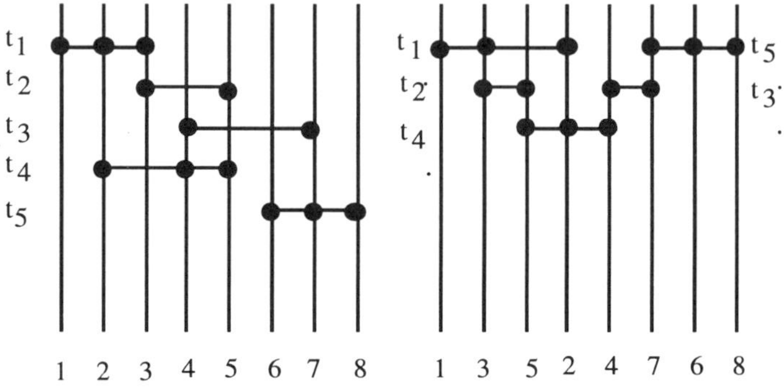

Fig. 1 Gates placement in 3 tracks.

A graph $G = (V, E)$ is an interval graph if there exists a set of finite closed intervals $\{ I(v) \mid v \in V \text{ and } (u, v) \in E \longleftrightarrow I(u) \cap I(v) \neq \phi \}$. The set of intervals is called a realization of the graph G. The set of vertices of the interval graph is the collection of nets. The vertices are adjacent iff the corresponding intervals intersect when they are placed on the same track. The intervals should be placed in tracks so that the required number of tracks is kept as small as possible. This is similar to the vertex coloring problem in graph theory. The necessary number of tracks is the *chromatic number* of the corresponding intervals graph. It is well known that the chromatic number of an interval graph is equal to the clique number or the maximum cardinality of all cliques. To solve the maximum clique problem we can use the Maximum Neural Network which resembles winner-takes-all circuit in a reasonable computation time. The maximum clique problem (cf. this volume) is first formulated as an unconstrained quadratic zero-one programming and it is solved by minimizing the weight summation over the same partition in a newly constructed graph.

2.3 GATES PACKING PROBLEMS

In one-dimensional logic arrays, the gates are arranged horizontally with each gate occupying a vertical strip. When the horizontal size of an array is fixed, the vertical spacing determines the total chip area which should be minimized. The vertical spacing is equivalent to the number of tracks. The existing algorithms provide a near-optimum ordering of the columns and determines the necessary minimum number of tracks. After the column assignment the gates must be placed on tracks. Our algorithm places and packs the gates on tracks within *O(1)*. Consider an example of 18 gates with 15 columns and 7 tracks given by Hong[1] in Fig. 2 :

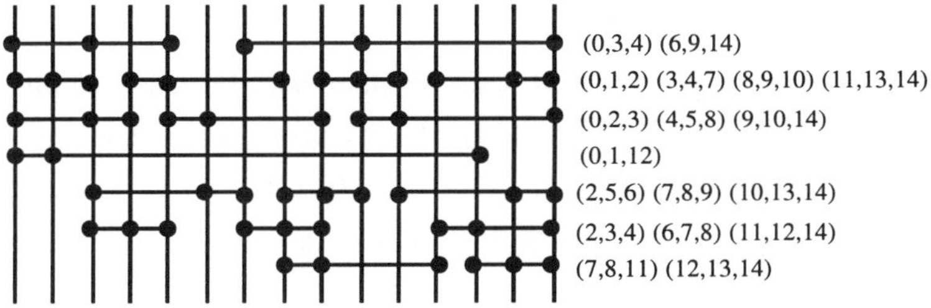

Fig. 2 18 gates placement in 15 columns and 7 tracks.

Note that (a,b,c) denotes a gate composed of three transistors located on a, b, and c column respectively. A 7 $\times$ 18 processor network array as shown in Fig. 3 is provided for packing the gates. Each gate will be placed on one of the seven tracks without any two gates crossing or overlapping each other.

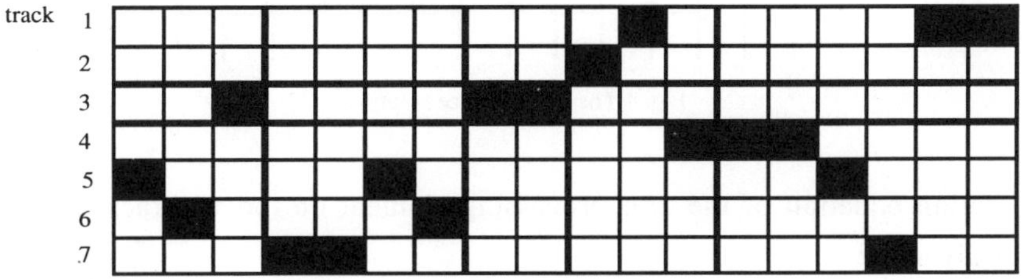

Fig. 3 a 7 $\times$ 18 neural network array.

* represents the initial random assignment of *i*th gate to *j*th track. The *i*th gate has a head number (the left-most column number) and a tail number (the right-most number) which are given by θ_i and τ_i respectively. All of the possible violation

conditions between two *i* and *j* gates on the same track can be addressed by θ_i and τ_i. These four violation conditions are given by the following functions:

$f(\theta_i, \theta_j, \tau_i) \times f(\theta_j, \tau_i, \tau_j)$ for (Case 1)
$f(\theta_i, \theta_j, \tau_i) \times f(\theta_j, \tau_j, \tau_i)$ for (Case 2)
$f(\theta_i, \tau_j, \tau_i) \times f(\theta_j, \theta_i, \tau_j)$ for (Case3)
$f(\theta_j, \theta_i, \tau_j) \times f(\theta_j, \tau_i, \tau_j)$ for (Case4)

where f(L, M, R) = 1 if L<=M <=R; 0 otherwise. Each net should satisfy the separation conditions, any two distinct nets must not violate the overlapping conditions as in the Channel Routing Problems}[5]. The convergence of the packing neural network to a solution for the 18-gate-7-track problem is shown in Fig. 4.

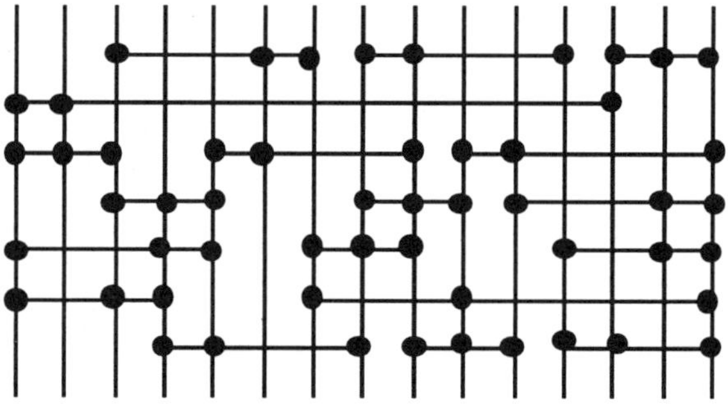

Fig. 4 The convergence solution.

The motion equation of the *xi*th processing element on the *x*th track for the *i*th gate is given by:

$$\frac{dU_{xi}}{dt} = -A\left(\sum_{k=1}^{m} V_{k,i} - 1\right) - Ch\left(\sum_{k=1}^{m} V_{k,i}\right) - B\sum_{j=1, j \neq i}^{n} V_{x,j}\, f(\theta_i, \theta_j, \tau_i) f(\theta_j, \tau_i, \tau_j) \quad (2)$$

$$- B \sum_{j=1, j \neq i}^{n} V_{x,j} f(\theta_i, \theta_j, \tau_i) f(\theta_j, \tau_j, \tau_i)$$

$$- B \sum_{j=1, j \neq i}^{n} V_{x,j} f(\theta_i, \tau_j, \tau_i) f(\theta_j, \theta_i, \tau_j)$$

$$- B \sum_{j=1, j \neq i}^{n} V_{x,j} f(\theta_j, \theta_i, \tau_j) f(\theta_j, \tau_i, \tau_j)$$

where h(x) is 1 if x = 0; 0 otherwise

In equation (2), A, B and C are constant. The first term forces one and only one *PE* to fire in *m*-track PE for the *i*th gate placement. The second term is a function for *Hill-Climbing* adjustment which enables the state of the system to escape from the local minimum. The remaining terms always perform the inhibitory forces as long as violations take place.

2.4 PARALLEL ALGORITHM

The parallel algorithm based on the concurrent computation network is simulated in C on a DEC 3100 workstation and in Turbo Pascal on a Macintosh SE/30. The following procedure describes the proposed parallel algorithm based on the *First Order Euler Method.*

1. Set t = 0, dt = 1, and A = B = C = 1.

2. The initial values of $U_{x,i}(t)$ for x = 1, ..., m and i = 1, ..., n are randomized.

3. Evaluate values of $V_{x,i}(t)$ in parallel based on the connectional binary function:

$$V_{x,i}(t) = \begin{cases} 1 & \text{if } U_{x,i}(t) > 0 \text{ and } U_{x,i}(t) = \max(U_{k,i}(t)) \text{ for } k = 1, \ldots, m. \\ 0 & \text{otherwise} \end{cases} \quad (4)$$

4.Use the motion equation in (2) to compute $dU_{x,i}(t)$ in parallel. v is the *heuristic function.*

$v = V_{x,i}(t)$ if (t mod 10) < 9

$= 1$ otherwise

$$dU_{xi} = -A(\sum_{k=1}^{m} V_{k,i} - 1) - Ch(\sum_{k=1}^{m} V_{k,i}) \quad (5)$$

$$- B\sum_{j=1, j \neq i}^{n} V_{x,j}\ f(\theta_i, \theta_j, \tau_i)f(\theta_j, \tau_i, \tau$$

$$- B\sum_{j=1, j \neq i}^{n} V_{x,j}\ f(\theta_i, \theta_j, \tau_i)f(\theta_j, \tau_j,$$

$$- B\sum_{j=1, j \neq i}^{n} V_{x,j}\ f(\theta_i, \tau_j, \tau_i)f(\theta_j, \theta_i,$$

$$- B\sum_{j=1, j \neq i}^{n} V_{x,j}\ f(\theta_j, \theta_i, \tau_j)f(\theta_j, \tau_i, \tau$$

where h(x) is 1 if x = 0; 0 otherwise

Cost function and temperature are not employed to accept or reject newly generated solutions. Instead, we use a simple periodical schedule function as v to generate all the possible solutions (perturbation mechanism). v and h(x) combine together are able to deal with the general problems of hill-climbing, such as local minimum, plateau, and ridge.

5. Compute $U_{x,i}(t=1)$ in parallel based on the first order Euler method:

$$U_{x,i}(t+1) = U_{x,i}(t) + dU_{x,i}/dt \quad \text{for } x = 1, ..., m \text{ and } i = 1, ..., n \quad (6)$$

6. Increment t by 1. If at least one from $dU_{1,i}(t)$ to $dU_{m,i}(t)$ is equal to zero for i = 1, ..., n then terminate this procedure else go to step 3.

2.5 SUMMARY

In this chapter, we presented an efficient parallel algorithm for gates packing problem. To validate our approach, the algorithm has been tested with the various examples posted by Hong[1] to justify the performance. More than a thousand simulation runs were performed to observe the frequency of the system to converge to the global minimum. The state of the system always converged to the global minimum. The average number of iterations for 18-gate-7-track problem is 2145. Table 1 show the statistics of the three problems. Figure 3-5 show the relationship between the frequency and the number of iteration steps to converge to the global minimum in the 48-gate-13-track problem. From the experimental results, we can conclude that the solution quality (speed of convergence) does not degrade with the problem size increase. We have demonstrated in this chapter that massive parallelism is inherent in gate packing problems. While we need to impose some constraints on the underlying model, the algorithm is general enough to produce useful results.

Table 1. Comparison of results.

gate	track	iteration	max	min
18	7	2145	9512	211
37	13	3015	8773	240
48	13	1154	9213	208

2.6 REFERENCES

[1] Hong, Y. S., Park, J. H., and Kim, M., "A Heuristic Algorithm for Ordering the Columns in One-Dimensional Logic Arrays," IEEE Trans. on CAD, Vol. 8, no. 5, pp. 547--562, Jun. 1987.

[2] Lopez, A. D. and Law, H., "A Dense Gate Matrix Layout Method for MOS VLSI," IEEE Trans. on Electron Devices, Vol. ED-27, pp. 1671--1675, 1980.

[3] Ohtsuki, T., Mori, H., Kuh, E. S., Kashiwabara T., and Fujisawa, T., "One-Dimensional logic Gate Placement and Interval Graphs," IEEE Trans. on Circuit System, Vol. CAS-26, pp. 675--684, Sep., 1979.

[4]Hashimoto, A. and Stenvens, J., "Wire Routing by Optimization Channel Assignment within Large Apertutrs," Proc. 18th Design Automation Workshop, pp. 155--168, 1971.

[5] Takefuji, Y., Neural Network Parallel Computing, chapter 5, Kluwer Academic Publishers, 1992.

[6] Weinberger, A., "Large Scale Intgration of MOS Complex Logic: A Layout Method," IEEE Journal of Solid-State Circuit, Vol. SC-2, pp. 182--190, Dec., 1967.

[7]Wing, O., Huang, S., and Wang, R., "Gate Matrix Layout," IEEE Trans. on CAD, Vol. CAD-6, pp. 79-84, Jan., 1987.

[8] Yamada, S., Okude, H., and Kasai, T., "A Hierarchical Algorithm for One-Dimensional Gate Assignment Based on Contraction of Nets," IEEE Trans. on CAD, Vol. 8, no. 6, pp.622-629, Jun., 1987.

Chapter 3

MAXIMUM CLIQUE PROBLEMS: PART 1

KUO-CHUN LEE
Cirrus Logic, Inc., R & D Department, Fremont, CA 94538, USA

YOSHIYASU TAKEFUJI
Dept. of Electrical Engineering and Applied Physics
Case Western Reserve University
Cleveland, OH 44106, USA
and
Faculty of Environmental Information
Keio University, Fujisawa 252, Japan

The maximum clique problem has been widely investigated in many research areas including cluster analysis, classification theory, graph coloring, information retrieval systems, disposal systems, VLSI circuit design, and biological systems. The maximum clique problem is to find the maximum complete subgraph of a given graph G. In the existing algorithms the branch-and-bound method has been most intensively studied for the maximum clique problem. The exponentially increasing computation time of the branch-and-bound method prohibits us to solve large scale real-world problems. A mathematical model fusing the Ising model and the artificial neural network for the maximum clique problem is proposed in this paper. A parallel algorithm based on the model is introduced where it solves large scale problems in reasonable computation time where the best known algorithms cannot solve. The maximum clique problem is first formulated as an unconstrained quadratic zero-one programming and it is solved by minimizing the weight summation over the same partition in a newly constructed graph. The theoretical and empirical convergence analysis are given where more than thirty thousand instances were simulated and tested. We compare our algorithm with two best known algorithms reported in 1990. First the computation time and the solution quality were compared between the best known algorithm on a supercomputer Cray 2 and our algorithm on a Macintosh SE/30. Note that the speed of the Cray 2 and that of the Macintosh SE/30 are 125 MIPS and 2 MIPS (million instructions per second) respectively. Secondly we compared our algorithm with a partially enumerative algorithm. As far as we have observed the behavior of the proposed system composed of N processing elements where N is the number of vertices, it converges within nearly O(1) time regardless of the problem size and it provides much the same solution quality and requires much less computation time than the best known algorithms.

3.1 INTRODUCTION

Many physico-chemical systems which encounter phase transitions can be represented by an array of lattice sites. This simple model is powerful enough to provide a unified theoretical basis for understanding a variety of phenomena such as ferromagnetism and antiferromagnetism, order/disorder transitions in alloys, etc. It is more convenient to describe those systems in the language of ferromagnetism [1]. Each node of lattice sites is regarded as an element with a magnetic moment (spin). In the Ising model, the spins are placed at the nodes of a crystalline lattice which is usually square or cubic. Each element is characterized by its spin, which takes the value of either +1 or -1, as a function of its interactions with its neighbors. Peierls [2] is the first one to demonstrate that at a sufficiently low temperature the Ising model in two or three dimension can show ferromagnetism. It is expected that the study of the Ising model in higher dimension would lead to understanding nature of a real physical system. The two-dimension spin glasses model has been solve by an efficient algorithm [3]. However, the computation complexity exponentially increases with higher dimension. Barahona [4] has shown that the problem of finding the ground state of three dimension spin glasses model is NP-complete [5]. In other words, no polynomial algorithm exists so far.

Given a graph G, a clique is a maximal complete subgraph of G. The maximum clique problem is to find the maximum complete subgraph of G. The problem is known to be NP-complete (Nondeterministic Polynomial) [5]. It is also well known that the maximum clique problem is very useful and it has been investigated in many areas including cluster analysis, classification theory, graph coloring, information retrieval systems, disposal systems, VLSI circuit design, and biological systems [6]. Many algorithms for the maximum clique problem have been reported [7-12]. In the existing algorithms the branch-and-bound method has been extensively studied. In 1990 Pardalos and Phillips [10] proposed an algorithm where the maximum clique problem is formulated as a linear constrained indefinite quadratical global optimization problem. Their algorithm is only able to solve the problem with less than 75 vertices although a supercomputer Cray 2 is used. Besides, the feasible solution is not always guaranteed by their algorithm whenever the system reaches the local minimum. In 1990 Pardalos and Rodger [11] transferred the maximum clique problem into an unconstrained quadratic zero-one programming problem where another branch-and-bound method is used in their algorithm. Their algorithm can deal with up to 1,000 vertices and 150,000 edges graph problems. Nevertheless, the performance of the branch-and-bound method drastically degrades as the problem size or the number of edges increases. A parallel algorithm for finding the maximum clique of a circular-arc graph was proposed by Kim in 1990 [13]. Unfortunately, no parallel algorithms for a general maximum clique problem have been reported.

It is believed that no polynomial time algorithm exists for the general maximum clique problem so far [5]. The near-optimum solution is practically acceptable as long as it can be obtained in a reasonable computation time. A recent research progress in the parallel/distributed system has motivated us to fully exploit parallelism using the artificial neural network for solving hard problems. The study of the artificial neural network system began in the early 1940s. McCulloch and Pitts showed that a neuron can be modeled as a simple threshold device which performs a binary logic function [14]. Later the neuron model was refined in Rosenblatt's Perceptron [15] and Widrow and Hoff's Adaline [16]. During 1970's, Grossberg and Carpenter [17] proposed an adaptive resonance theory neural network based on the assumption that the brain spontaneously organizes itself into recognition codes. Meanwhile, Kohonen developed a self-organizing process by which an optimal topographical mapping of the signal is constructed [18]. Recently, Rumelhart and McClelland [19] proposed the backpropagation algorithm based on the feedforward multi-layered neural network. In 1985 Hopfield and Tank [20] used an artificial neural network to solve combinatorial optimization problems such as travelling salesman problem. In Hopfield neural network, the parallel gradient descent method seeks the local minimum of the given Liapunov energy function E. Generally, the Liapunov energy function E is given by the necessary and sufficient constraints and/or the cost function. The synaptic weights between neurons are determined by the energy function E. In general, the state of the system in the Hopfield neural network is guaranteed to converge to one of the local minima instead of the global minimum [21]. However, the local minimum is not always equivalent to the feasible solution. In other words, the proof of the local minimum convergence in the Hopfield neural network cannot always guarantee a acceptable solution. In this paper a parallel algorithm based on the maximum neural network is proposed for the maximum clique problem. The maximum neural network is a mathematical model fusing the Ising model and the artificial neural network. The convergence properties of the maximum neural network and the relationship between local minima and feasible solutions are given. A massive number of simulation runs were performed for testing the proposed algorithm and evaluating the solution quality and the computation time with two best known algorithms

3.2. MODELS OF THE PROCESSING ELEMENT

The artificial neural network consists of a massive number of simple processing elements where they are highly interconnected. The processing element is called the artificial neuron, because it performs the simplified function of a biological neuron. The mathematical model of the artificial neural network is composed of two important components: neurons and synaptic links. The output signal generated by one neuron propagates to the others through the synaptic links. The new state of the neuron is determined by the linear sum of the weighted input signals. McCulloch and Pitts [14] proposed a mathematical neuron model based on biological computation in 1943. The input/output function of the McCulloch-Pitts neuron is given by $V_i = f(U_i) = 1$ if $U_i \geq 0$ and 0 otherwise where V_i and U_i are the output and the input of the *i*th neuron respectively. The artificial neural network for solving combinatorial optimization problems was first introduced by Hopfield and Tank in 1985 [20] where they used the sigmoid neuron. Hopfield and Tank mapped the travelling salesman problem onto an N x N artificial neural network. The energy function E for an N-city problem is given by

$$E = \frac{A}{2}\sum_{x}^{N}\sum_{i}^{N}\sum_{j\neq i}^{N} V_{xi}V_{xj} + \frac{B}{2}\sum_{i}^{N}\sum_{x}^{N}\sum_{x\neq y}^{N} V_{xi}V_{yi} + \frac{C}{2}\left(\left(\sum_{i}^{N}\sum_{x}^{N} V_{xi} - N\right)\right)^2$$

$$+ \frac{D}{2}\sum_{x}^{N}\sum_{y\neq x}^{N}\sum_{i}^{N} d_{xy}V_{xi}(V_{y,i+1} + V_{y} \qquad (1)$$

where d_{xy} is the distance between city x and city y.

The dynamics of the system follows the motion equation which is given by

$$\frac{dU_{Xi}}{dt} = -\frac{U_{Xi}}{\tau} - \frac{\partial E}{\partial V_{Xi}} \quad \text{for X, i =1 to N} \tag{2}$$

It has been shown that the decay term $-\frac{U_{Xi}}{\tau}$ has an undesirable effect for the system convergence [22]. Because it may increase the energy function $E(V_{1,1}, \ldots, V_{N,N})$ in Eq.(1) under some conditions. In other words, E is not the Liapunov energy function of the system if the decay term is involved in the motion equation. Many researchers use a large value for τ in order to eliminate the undesirable effect. It is suggested that the decay term should be removed from the motion equation based on both theoretical and empirical reasons [22]. The theoretical detail is shown in [22]. The McCulloch-Pitts neuron model and the modified McCulloch-Pitts neuron model have been successfully used (without the decay term in the motion equation) for graph planarization problems [23], tiling problems [24], RNA secondary structure prediction problems [25][26], maximum independent set problems [26], sorting problems [27], k-colorability problems [22], and spare allocation problems [28].

In this paper a maximum neural network is introduced for the maximum clique problem. Inspired by the Ising model and the parallelism in the artificial neural network, the maximum neural network is invented. The maximum neural network consists of *M* clusters where each cluster is composed of *n* processing elements. The total number of required processing elements is *M* x *n*. The input/output function of the *i*th maximum neuron in the *m*th cluster is given by : $V_{m,i} = 1$ if $U_{m,i} = \max\{U_{m,1}, \ldots, U_{m,n}\}$, 0 otherwise. Therefore one and only one processing element among n processing elements per cluster is always encouraged to fire[1] in the maximum neural network. The maximum neural network model has been successfully used to solve the tiling problem [24], the max cut problem [29], the bipartite subgraph problem [30], and the module orientation problem [31].

---

[1] "Firing a neuron" means that the output of the neuron becomes one.

3.3 MAXIMUM CLIQUE PROBLEMS

The unconstrained quadratic zero-one programming problem has been investigated in many research areas including financial analysis problems [32], CAD problems [33], and models of message traffic management [34]. The problem is formulated by:

$$\text{minimize } f(x) = C^T x + x^T Q x, \qquad x_i \in \{0,1\} \tag{3}$$

where Q is an N x N symmetrical rational matrix with zero diagonal elements and C is a rational N-vector. Eq.(3) is rewritten in the following form

$$f(x) = \sum_{i=1}^{N} c_i x_i + \sum_{i=1}^{N} \sum_{j=1}^{N} q_{i,j} x_i x_j, \qquad x_i \in \{0,1\} \tag{4}$$

where $q_{i,j}$ is the *ij*th element of Q and c_i is the *i*th element of the column vector C.

It is known that many graph problems can be formulated as an unconstrained quadratic zero-one programming problem [7]. Given a graph G(V,E) where V is a set of vertices and E is a set of edges, a complete subgraph of G is one whose vertices are pairwise adjacent. A clique is a maximal complete subgraph of G. The maximum clique problem is to find the maximum complete subgraph of G. When $A_{\bar{G}}$ is the adjacency matrix of the complement of G, and I is an identity matrix, minimizing the following unconstrained quadratic zero-one programming problem is equivalent to finding a maximum clique in G [7][11]:

$$\text{To minimize the function } f(x) = x^T(A_{\bar{G}} - I)x \text{ for } x_i \in \{0,1\} \tag{5}$$

is equivalent to minimizing

$$f(x) = -\sum_{i=1}^{N} x_i + \sum_{i=1}^{N}\sum_{j=1}^{N} q_{i,j}\, x_i x_j \text{ for } x_i \in \{0,1\} \tag{6}$$

where $q_{i,j} = 1$ if there exists an edge between the *i*th vertex and the *j*th vertex in the graph $\overline{G}=(V,\overline{E})$, and $q_{i,j} = 0$ otherwise.

The quadratic problem belongs to one of certain NP-complete problems. Transformation from the maximum clique problem in a quadratic zero-one form to minimizing the weights in a newly construct graph G_M is shown in Appendix A. The equivalence between the quadratic zero-one programming problem and the maximum clique problem has been reported in [35][36]. The goal of the maximum clique problem is finally formulated by minimizing the following function E:

$$E = \sum_{x=0}^{N}\sum_{y=0}^{N}\sum_{i=1}^{2} W(x,y)\, V_{x,i}\, V_{y,i} \tag{7}$$

$$\text{subject to } \sum_{i=1}^{2} V_{x,i} = 1 \text{ for } x = 0 \text{ to } N$$

$$\text{for } V_{x,i} \in \{1, 0\} \text{ and } V_{0,1} = 1$$

In other words, finding the maximum clique in a graph G is equivalent to minimizing the weight summation over the same partition in graph G_M.

3.4 MAXIMUM NEURAL NETWORK

The existing neural networks for solving optimization problems are based on the Liapunov energy function E. An energy function is composed of two parts. One is to keep the state of the network in a valid solution. The other part represents the solution cost. The sum of these two parts forms the computational energy function. For example, the first three terms in Eq.(1) for the travelling salesman problem represent the validity of the solution and the last term represents the cost. The state of the local minimum is not always guaranteed to be a valid solution. This is why the valid tour for the travelling salesman problem cannot be always obtained even in the small size problem. On the other hand, we must suffer from determining the coefficient parameters in the computational energy function E. No theoretical analysis has been provided for generally determining the coefficient parameters.

In the maximum neural network, the input/output function of each neuron is given by : $V_{x,i}$ =1 if $U_{x,i}$ is the maximum value between $V_{x,1}$ and $V_{x,2}$, 0 otherwise[2] where x and i represent the vertex number and partition number respectively. Due to this function, there is always one and only one neuron fired for the *x*th vertex. The maximum clique problem with N vertices can be solved by minimizing the following energy function E

$$E = \frac{C}{2} \sum_{x=0}^{N} \sum_{y=0}^{N} \sum_{i=1}^{2} W(x,y)\, V_{x,i}\, V_{y,i} \tag{8}$$

The motion equation for the *xi*th neuron is given by

$$\frac{dU_{x,i}}{dt} = -\sum_{y=0}^{N} W(x,y)\, V_{y,i} \quad \text{for } C=1 \tag{9}$$

It is significantly advantageous that the coefficient parameter tuning is not required in the maximum neural network for solving the maximum clique problem.

3.5 CONVERGENCE PROPERTY

Three theorems are given in the Appendix B. Theorem 1 guarantees the local minimum convergence of the maximum neural network. Theorem 2 states that the solution quality improves as time elapses until no further improvement can be achieved. Theorem 3 states that as long as the state of the system converges to the

---

[2] If $U_{x,1} = U_{x,2}$, then the network will set $V_{x,1}=1$ and $V_{x,2}=0$.

local/global minimum, the final converged state always corresponds to a maximal/maximum clique. Our parallel algorithm is based on the maximum neural network where it not only runs on a sequential machine but also on a parallel machine. The termination condition of our algorithm is given by detecting the convergence state. As long as the system reaches a stable point or an equilibrium state, the algorithm is terminated. The equilibrium state is defined that all firing neurons have the smallest change rate of the input per cluster. The condition of the equilibrium state is given by:

$$V_{x,i}(t) = 1 \text{ and } \frac{dU_{x,i}(t)}{dt} = \min\left\{\frac{dU_{x,1}(t)}{dt}, \frac{dU_{x,2}(t)}{dt}\right\} \text{ for } x=1 \text{ to } N. \tag{10}$$

In summary, the maximum neural network has the following advantages for solving the maximum clique problem and certain NP-complete problems: 1) coefficient-parameter tuning in the motion equation is not required in the maximum neural network while the conventional neural networks must suffer from it; 2) the equilibrium state of the maximum neural network is clearly defined in order to terminate the algorithm while the existing neural networks do not have the clear definition; and 3) the maximum neural network always allows the state of the system to converge to the feasible solution while the existing neural networks cannot guarantee it.

3.6 PARALLEL ALGORITHM AND SIMULATION

The parallel algorithm based on the maximum neural network is implemented in C on a DEC3100 machine and in Turbo Pascal on a Macintosh SE/30. The following procedure describes the proposed parallel algorithm based on the first order Euler method.

step 0: Set $t=0$

step 1: **for** $x=1$ **to** N **do**

for $i=1$ **to** 2 **do in parallel**

$U_{x,i}(t) \leftarrow$ Random number $\in (-1,1)$

step 2: **for** $x=1$ **to** N **do**

for $i=1$ **to** 2 **do in parallel**

$V_{x,i}(t) = 1$ if $U_{x,i}(t) = \max\{U_{x,1}(t), U_{x,2}(t)\}$

0 otherwise

step 3: **for** $x=1$ **to** N **do**

for $i=1$ **to** 2 **do in parallel**

$$\Delta U_{x,i}(t) = -\sum_{y=0}^{N} W_{x,y} V_{y,i}$$

step 4: **for** $x=1$ **to** N **do**

for $i=1$ **to** 2 **do in parallel**

$U_{x,i}(t+1) = U_{x,i}(t) + \Delta U_{x,i}(t)$

step 5: Increment t by 1.

step 6: If the state of the system reaches an equilibrium state then stop else goto step 2.

In the proposed parallel algorithm, the initial state is randomly generated for the input of each processing element at step 1. At step 2, the output of each processing element is evaluated according to the maximum function. The motion equation in Eq.(9) determines the change of each input of the processing elements at step 3. The new state of each input is updated by the first order Euler method at step 4. The solution is interpreted by the final state of the system. If $V_{x,1} = 1$, then the xth vertex belongs to the clique. Therefore the total number of vertices which belong to the first partition is the maximum/maximal clique size. A 10-vertex graph is given below as an example to explain our algorithm.

Example: A 10-vertex graph is given in Fig.1(a) and its complementary graph $\overline{G}$ is shown in Fig.1(b). The first step is to construct a graph G_M based on Eq.(A.3) in Appendix A. G_M is constructed by adding a vertex 0 and ten edges from vertex 0 to other vertices in graph $\overline{G}$. The graph G_M is shown in Fig.1(c) and its weight matrix W is given in Fig.1(d). The weight matrix can be computed by $W_{i,j} = W_{j,i} = \frac{1}{4} q_{i,j}$, and $W_{0,i} = W_{i,0} = \frac{1}{4}\left(\sum_{j=1}^{N} q_{i,j} - 1\right)$ for i, j = 1 to 1(where $q_{i,j}$ is 1 if there is an edge between the *i*th vertex and the *j*th vertex in graph $\overline{G}$, 0 otherwise. For example $W_{2,4}$ =0.25 is given because there is an edge between the second vertex and the forth vertex in graph $\overline{G}$. The second step is to apply the proposed parallel algorithm to solve the maximum clique problem. One of the solutions is shown in Fig.1(e) where the black color and grey color represent two different partitions respectively. The vertices with black color form a maximum/maximal clique in graph G.

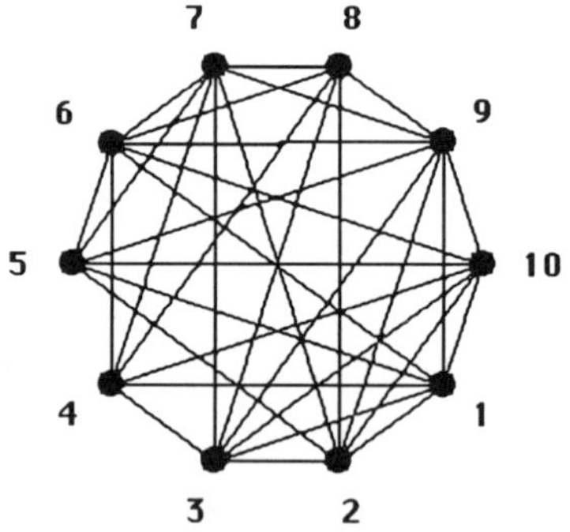

Fig. 1(a) A 10-vertex graph G with 34 edges.

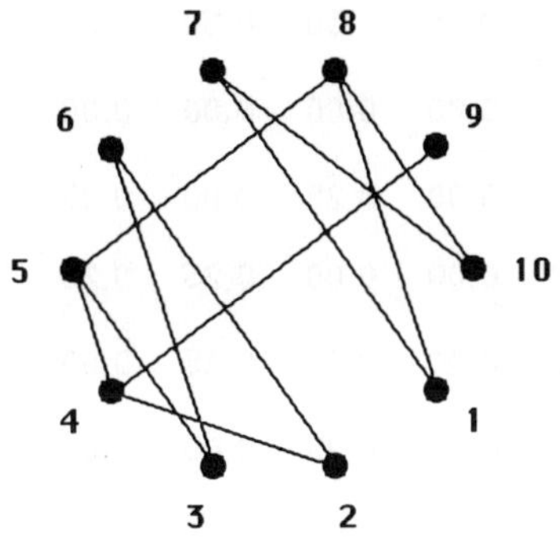

Fig. 1(b) Complementary graph of G.

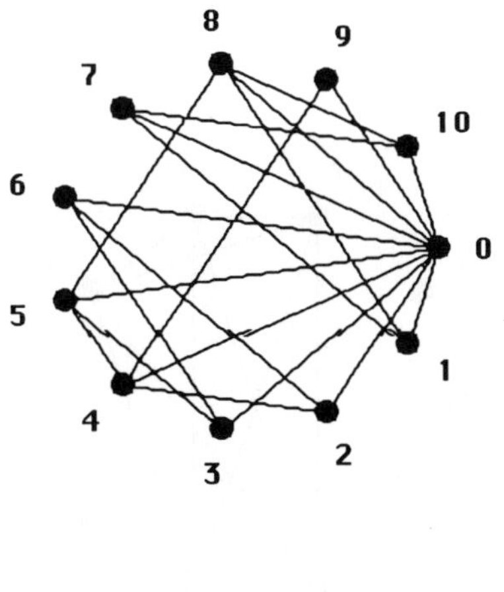

Fig. 1(c) The graph G_M

	0	1	2	3	4	5	6	7	8	9	10
0	0.00	0.25	0.25	0.25	0.50	0.50	0.25	0.25	0.50	0.00	0.25
1	0.25	0.00	0.00	0.00	0.00	0.00	0.00	0.25	0.25	0.00	0.00
2	0.25	0.00	0.00	0.00	0.25	0.00	0.25	0.00	0.00	0.00	0.00
3	0.25	0.00	0.00	0.00	0.00	0.25	0.25	0.00	0.00	0.00	0.00
4	0.50	0.00	0.25	0.00	0.00	0.25	0.00	0.00	0.00	0.25	0.00
5	0.50	0.00	0.00	0.25	0.25	0.00	0.00	0.00	0.25	0.00	0.00
6	0.25	0.00	0.25	0.25	0.00	0.00	0.00	0.00	0.00	0.00	0.00
7	0.25	0.25	0.00	0.00	0.00	0.00	0.00	0.00	0.00	0.00	0.25
8	0.50	0.25	0.00	0.00	0.00	0.25	0.00	0.00	0.00	0.00	0.25
9	0.00	0.00	0.00	0.00	0.25	0.00	0.00	0.00	0.00	0.00	0.00
10	0.25	0.00	0.00	0.00	0.00	0.00	0.00	0.25	0.25	0.00	0.00

Fig. 1(d) The weight matrix of graph G_M.

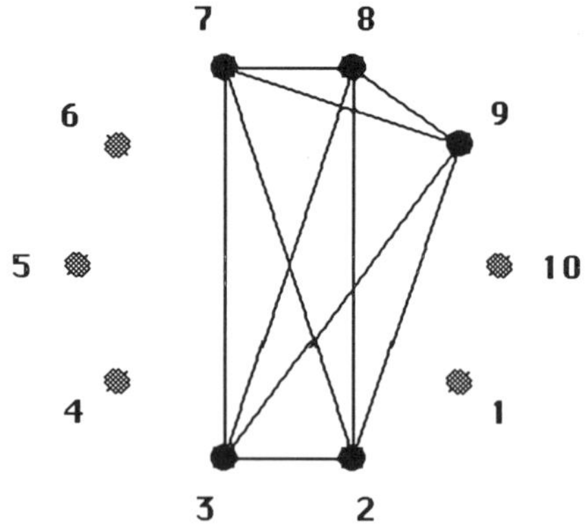

Fig. 1(e) The solution for the graph in Fig.1(a).

In order to evaluate the solution quality and the computation time of our parallel algorithm, it is first compared with the latest algorithm proposed by Pardalos and Philips [10] where a graph with 10 vertices to 100 vertices and with density from d=0.1 to d=0.9 was tested on the supercomputer Cray 2 as shown in Table 1-5. For a given N-vertex graph, the density of the problem is defined by the ratio between the number of given edges and N(N-1)/2. Our simulator has been developed in Turbo Pascal on a Macintosh SE/30 machine. The comparisons of the simulation results are shown in Table 1-5.

Table 1. Performance comparisons between Pardalos and Philips' algorithm in [10] on a supercomputer Cray 2 and our algorithm on a Macintosh SE/30 on average density d_{ave}, maximum clique size k_{ave}, and CPU time t_{ave} (seconds) for 10 graphs where each graph has 10 vertices.

Cray 2			Macintosh SE/30		
d_{ave}	k_{ave}	t_{ave}	d_{ave}	k_{ave}	t_{ave}
0.10	2.1	0.3	0.09	1.9	0.4
0.22	2.9	0.5	0.21	2.0	0.5
0.50	3.7	0.6	0.51	3.9	0.4
0.75	5.5	0.4	0.73	4.5	0.4
0.88	7.0	0.2	0.89	7.1	0.4

Table 2. Performance comparisons between Pardalos and Philips' algorithm in [10] on a supercomputer Cray 2 and our algorithm on a Macintosh SE/30 on average density d_{ave}, maximum clique size k_{ave}, and CPU time t_{ave} (seconds) for 10 graphs where each graph has 25 vertices.

Cray 2			Macintosh SE/30		
d_{ave}	k_{ave}	t_{ave}	d_{ave}	k_{ave}	t_{ave}
0.09	2.6	6.7	0.10	2.2	1.9
0.24	3.4	7.3	0.25	2.8	2.1
0.49	3.4	7.3	0.49	4.9	1.9
0.75	8.7	6.7	0.73	7.7	1.8
0.91	14.2	4.7	0.89	12.8	1.8

Table 3. Performance comparisons between Pardalos and Philips' algorithm in [10] on a supercomputer Cray 2 and our algorithm on a Macintosh SE/30 on average density d_{ave}, maximum clique size k_{ave}, and CPU time t_{ave} (seconds) for 10 graphs where each graph has 50 vertices.

Cray 2			Macintosh SE/30		
d_{ave}	k_{ave}	t_{ave}	d_{ave}	k_{ave}	t_{ave}
0.10	2.7	93.6	0.10	2.7	9.3
0.25	3.6	98.1	0.25	3.9	8.9
0.50	5.9	86.3	0.49	5.8	7.8
0.75	10.4	78.7	0.74	10.8	6.9
0.90	18.8	69.2	0.89	17.9	6.4

Table 4. Performance comparisons between Pardalos and Philips' algorithm in [10] on a supercomputer Cray 2 and our algorithm on a Macintosh SE/30 on average density d_{ave}, maximum clique size k_{ave}, and CPU time t_{ave} (seconds) for 10 graphs with vertex size 75

Cray 2			Macintosh SE/30		
d	k	t	d_{ave}	k_{ave}	t_{ave}
0.09	3.0	2283.0	0.10	2.9	26.2
0.24	4.0	3690.1	0.25	4.3	24.3
0.49	6.0	2837.1	0.50	6.4	20.2
0.75	11.0	2941.4	0.75	12.7	16.5
0.91	22.0	3879.6	0.90	22.2	14.7

Table 5. Performance of our algorithm on a Macintosh SE/30 on average density d_{ave}, maximum clique size k_{ave}, and CPU time t_{ave} (seconds) for 10 graphs where each graph has 100 vertices.

d_{ave}	k_{ave}	t_{ave}
0.10	3.0	51.0
0.25	4.4	45.8
0.50	7.0	38.3
0.75	12.4	30.4
0.89	24.1	26.5

Because the computation time of Pardalos and Philips' algorithm on Cray 2 drastically increases, only one 75-vertex problem was reported in [10]. It is obvious that the solution qualities of both algorithms are much the same and our algorithm outperforms Pardalos and Philips' algorithm in terms of the computation speed. Remember that the speed of the Cray 2 is 125 MIPS and that of the Macintosh SE/30 is 2 MIPS. Our algorithm is faster than their algorithm at least 600 times in Table 3 and 6000 times in Table 4 with much the same solution quality. As Pardalos and Philips pointed out in their paper [10], it is not recommended to use their algorithm for greater than 75-vertex problem because of the prohibitively long computation time. To substantiate our algorithm, the simulation result for a graph with 100-vertex size is also shown in Table 5.

It is expected that the computation time of the branch-and-bound method drastically increases as the density of the given problem increases even when the problem size is fixed. Because the number of search tree nodes increases as the density increases. However, our algorithm behaves totally different. For higher density, the computation time becomes shorter. Because the graph G_M is obtained by adding a node s_0 to the graph $\overline{G}$ which is the complementary graph of the given graph.

Table 6. The CPU time (seconds) of our algorithm for the graph size ranging from 500-vertex to 1000-vertex with densities from 0.5 to 0.9 on a DEC3100 machine (13 MIPS).

	Density				
Size	**0.5**	**0.6**	**0.7**	**0.8**	**0.9**
500	141.7	135.7	131.2	155.4	332.1
600	187.8	175.6	200.0	516.4	222.5
700	311.4	289.1	230.7	267.6	358.7
800	460.8	539.9	852.4	465.5	835.7
900	1008.9	524.6	509.8	379.2	1082.5
1000	1488.1	695.3	687.2	858.2	1337.1

Our algorithm is also compared with a partially enumerative algorithm presented by Carraghan and Pardalos in 1990 [12] which is faster than any previously known existing algorithms. Their simulation results were obtained on an IBM 3090 computer. The speed of the IBM3090 is 30 MIPS. According to their results, the CPU time for a 500-vertex graph with 0.5 density is more than 1000 seconds while the CPU time becomes prohibitively long even for a 200-vertex graph with 0.9

density. Carraghan and Pardalos' algorithm could not solve the graph with more than 400 vertices and more than 0.6 density in reasonable time. The simulation result is compared in Appendix C where our algorithm generates solutions with much the same quality. Our algorithm was also tested for unsolved graph problems with more than 400 vertices and more than 0.6 density. The simulation result is summarized in Table 6. It shows that our algorithm is able to solve large unsolved problems in reasonable time. A graph with 1000-vertex and 0.9 density has 1,998,000 edges.

3.7 CONCLUSION

In this paper a new computation model for large scale maximum clique problems is proposed and tested. The proposed maximum neural network has the following advantages: 1) coefficient-parameter tuning in the motion equation is not required in the maximum neural network while the conventional neural networks must suffer from it; 2) the equilibrium state of the maximum neural network is clearly defined in order to terminate the algorithm while the existing neural networks do not have the clear definition; and 3) the maximum neural network always allows the state of the system to converge to the feasible solution while the existing neural networks cannot guarantee it. The proposed parallel algorithm for large size problems outperforms over the best known algorithms in terms of the computation time with much the same solution quality where the conventional branch-and-bound method cannot be used due to the exponentially increasing computation time. The equivalence of certain NP-complete problems is shown in Fig.2 where $A \leftrightarrows B$ means that A and B are mathematically identical and $A \rightarrow B$ means that B is a subset of A. This new computation model can be used not only to solve every problem in Fig.2 without any transformation but also more than 300 applications with polynomial transformation [5]. The massive simulation results indicate that the number of iteration steps to converge to the solution does not increase with the problem size. It is significantly encouraging to designing and building a parallel machine based on our algorithm.

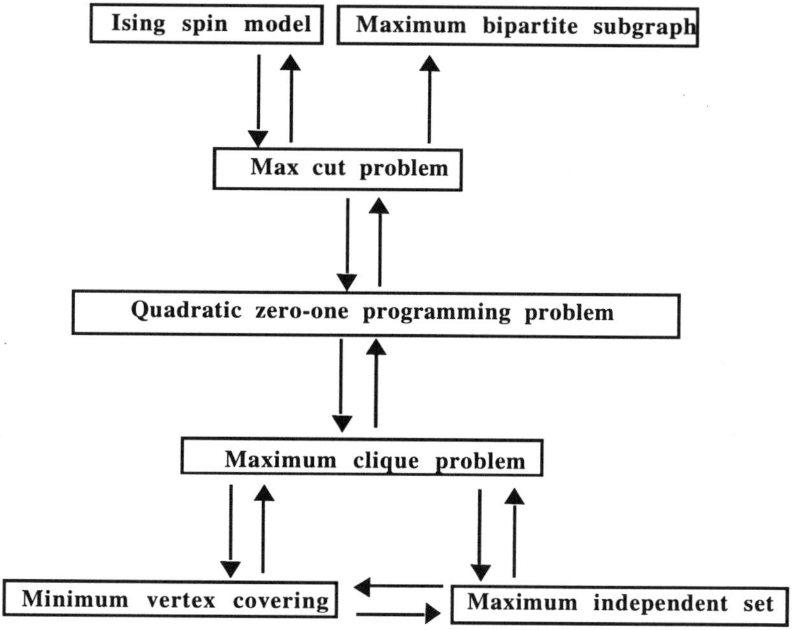

Fig. 2 The relationship between certain NP-complete problems.

3.8 REFERENCES

[1] P. K. Pathria, Statistical Mechanics, New York: Pergamon Press Inc., 1972.

[2] R. E. Peierls, Proc. Camb. Phil. Soc. 32, 471, 477, 1936,

[3] I. Bieche, R. Maynard, R. Rammal and J. P. Uhry, J. Phys. A: Math. Gen. 13, pp. 2553-2576, 1980.

[4] F. Barahona, R. Maynard, R. Rammal and J. P. Uhry, J. Phys. A: Math. Gen. 15, pp. 673-699, 1982.

[5] M. R. Garey and D. S. Johnson, Computers and Intractability : A Guide to the Theory of NP-Completeness, New York: W. H. Freeman and Co., 1979.

[6] L. Gerhards and W. Lindenberg, "Clique detection for nondirected graphs: two new algorithms," Computing, 21, pp. 295-322, 1979.

[7] P. M. Pardalos and G. P. Rodgers, "Computational aspects of a branch and bound algorithm for quadratic zero-one programming," Computing, 45, pp. 131-144, 1990.

[8] E. Balas and C. S. Yu, "Finding a maximum clique in a arbitrary graph," SIAM J. Comput., vol. 15, no. 4, pp. 1054-1068, 1986.

[9] E. Bron and J. Kerbosch, "Finding all cliques of an undirected graph," Communication of the ACM, 16, no. 9, pp. 575-577, 1973.

[10] P. M. Pardalos and A. T. Phillips, "A global optimization approach for solving the maximum clique problem," International J. Computer Mathematics, vol. 33, pp. 209-216, 1990.

[11] P. M. Pardalos and G. P. Rodgers, "A branch and bound algorithm for the maximum clique problem," To appear in Mathematical Programming.

[12] R. Carraghan and P. M. Pardalos, "An exact algorithm for the maximum clique problem," Operations Research Letters, 9, pp. 375-382, 1990.

[13] S. K. Kim, "A parallel algorithm for finding a maximum clique of a set of circular arcs of a circle," Information Processing Letters, 34, pp. 235-241, 1990.

[14] W. S. McCulloch, and W. H. Pitts, "A logical calculus of ideas immanent in nervous activity," Bulletin of Mathematical Biophysics, vol. 5, p. 115, 1943.

[15] F. Rosenblatt, "The perceptron: a probabilistic model for information storage

and organization in the brain," Psychological Review, 65, pp. 386-408, 1958.

[16] B. Widrow and M. E. Hoff, "Adaptive switching circuits," 1960 IRE Convention Record : Part 4- Computers; Man-Machine Systems, Los Angels, PP. 96-104, 1960.

[17] G. Carpenter and S. Grossberg, "A massively parallel architecture for a self-organization neural pattern recognition machine," Comput. Vision, Graph. Image Processing, vol. 37, pp. 54-115, 1987.

[18] T. Kohonen, Self-Organization and Associated Memory. New York: Springer-Verlag, 1984.

[19] D. E. Rumelhart and J. L. McClelland, Parallel Distributed Processing, Vol I: Foundations. Cambridge, MA: MIT Press, 1986.

[20] J. J. Hopfield and D. W. Tank, "Neural computation of decisions in optimization problems," Biological Cybernetics, vol. 52, pp. 141-152, 1985.

[21] J. J. Hopfield, "Neurons with graded response have collective computational properties like those of two-state neurons", Proc. Natl. Acad. Sci. USA, 81, pp. 3088-3092, 1984.

[22] Y. Takefuji, and K. C. Lee, "Artificial neural networks for four-coloring map problems and k-colorability problems," IEEE Trans. on Circuits and Systems, vo. 38, no. 3, pp. 326-333, 1991.

[23] Y. Takefuji and K. C. Lee, "A near-optimum parallel planarization algorithm," Science, vol. 245, pp. 1221-1223, Sept. 1989.

[24] Y. Takefuji and K. C. Lee, "A parallel algorithm for tiling problems," IEEE Trans. on Neural Networks, vol. 1, no. 1, pp. 143-145, March 1990.

[25] Y. Takefuji, C. W. Lin, and K. C. Lee, "A parallel algorithm for estimating the secondary structure in Ribonucleic Acids," Biological Cybernetics, 63, 5, pp. 337-340, 1990.

[26] Y. Takefuji, Li-lin Chen, K. C. Lee and J. Huffman, "Parallel algorithm for finding a near-maximum independent set of a circle graph," IEEE Trans. on Neural Networks, vol. 1 , no. 3, pp. 263-267, Sep. 1990.

[27] Y. Takefuji and K. C. Lee, "A super parallel sorting algorithm based on neural

networks" IEEE Trans. on Circuits and Systems, vol. 37, 11, pp. 1425-1429, 1990.

[28] N. Funabiki and Y. Takefuji, "A parallel algorithm for spare allocation problems ," To appear in IEEE Trans. on Reliability.

[29] K. C. Lee and Y. Takefuji, "A parallel algorithm using the maximum neural network for the max cut problem," Case Western Reserve University, CAISR Tech. Rep. TR91-104.

[30] K. C. Lee, Y. Takefuji et al., "A parallel improvement algorithm for the bipartite subgraph problem," Case Western Reserve University, CAISR Tech. Rep. TR91-105.

[31] K. C. Lee, Y. Takefuji et al., "A generalized maximum neural network for the module orientation problem," Case Western Reserve University, CAISR Tech. Rep. TR91-106.

[32] R. D. McBride and J. S. Yormark, "An implicit enumeration algorithm for quadratic integer programming," Management Science, 26, 3, pp. 131-143, 1980.

[33] J. Krarup and P. A. Pruzan, "Computer aided layout design," Mathematical Programming Study, 9, pp. 75-94, 1978.

[34] G. Gallo, P. L. Hammer, and B. Simeone, "Quadratic knapsack problems," Mathematical Programming, 12, pp. 132-149, 1980.

[35] P. L. Hammer, "Some network flow problems solved with pseudo-boolean programming," Operations Research, 13, pp. 388-399, 1965.

[36] F. Barahona, M. Junger, and G. Reinelt, "Experiments in quadratic 0-1 programming," Mathematical Programming, 44, pp. 127-137, 1989.

[37] M. L. Yu, " A study of the applicability of Hopfield decision neural nets to VLSI CAD," Proc. 26th Design Automation Conference, pp. 412-417, 1989.

Appendix A:

The goal of the quadratic zero-one programming problem is to minimize the following equation.

$$f(x) = \sum_{i=1}^{N} c_i x_i + \sum_{i=1}^{N}\sum_{j=1}^{N} q_{i,j}\, x_i x_j \text{ for } x_i \in \{0,1\} \tag{A.1}$$

Let $s_i = 2x_i - 1$, then Eq.(A.1) is equivalent to minimizing

$$\begin{aligned} f(x(s)) &= \sum_{i=1}^{N}\sum_{j=1}^{N} \frac{1}{4} q_{i,j} s_i s_j + \sum_{i=1}^{N}\sum_{j=1}^{N} \frac{1}{4} q_{i,j} s_i \\ &\quad + \sum_{i=1}^{N}\sum_{j=1}^{N} \frac{1}{4} q_{i,j} s_j + \sum_{i=1}^{N} \frac{1}{2} c_i s_i + C_1 \\ &= \sum_{i=1}^{N}\sum_{j=1}^{N} \frac{1}{4} q_{i,j} s_i s_j + \sum_{i=1}^{N}\sum_{j=1}^{N} \frac{1}{4} q_{i,j} s_i + \sum_{j=1}^{N}\sum_{i=1}^{N} \frac{1}{4} q_{j,i} s_i + \sum_{i=1}^{N} \frac{1}{2} c_i s_i + C_1 \\ &= \sum_{i=1}^{N}\sum_{j=1}^{N} \frac{1}{4} q_{i,j}\, s_i s_j + \sum_{i=1}^{N} \left(\sum_{j=1}^{N} \frac{1}{2} q_{i,j} + \frac{1}{2} c_i\right) s_i + C_1 \end{aligned} \tag{A.2}$$

for $s_i \in \{+1, -1\}$ and $C_1 = \sum_{i=1}^{N}\sum_{j=1}^{N} \frac{1}{4} q_{i,j} + \sum_{i=1}^{N} \frac{1}{2} c_i$.

Let $W_{0,i} = W_{i,0} = \frac{1}{4}\left(\sum_{j=1}^{N} q_{i,j} + c_i\right)$ for i = 1 to N and

$W_{i,j} = W_{j,i} = \frac{1}{4} q_{i,j}$ for i, j = 1 to N, and adding a variable $s_0 = 1$ (A.3)

then Eq.(A.1) is given by

$$f(s) = \sum_{i=0}^{N}\sum_{j=0}^{N} W_{i,j}\, s_i s_j + C_1 \text{ with } s_0 = 1 \text{ and } s_i \in \{ +1, -1 \} \tag{A.4}$$

A graph G_M=(V,E) with a vertex set V={0, 1, 2, ..., N} and an edge set E is constructed where $W_{i,j}$ defined in Eq.(A.3) represents an edge weight between the

xth vertex and the yth vertex. The assignment of each s_i, either s_i=+1 or s_i=-1, corresponds to a partition of V into $V^+=\{i\in V|\ s_i=+1\}$ and $V^-=\{i\in V|\ s_i=-1\}$. Note that the *0*th vertex has been assigned to the partition V^+. Because C_1 is a constant parameter, minimizing Eq.(A.1) is equivalent to minimizing

$$f(s) = \sum_{s_i,\, s_j \in V^+} W_{i,j} + \sum_{s_i,\, s_j \in V^-} W_{i,j} - \sum_{s_i \in V^+ \text{and } s_j \in V^-} W_{i,j} + C_1$$

$$= 2\cdot\left(\sum_{s_i,\, s_j \in V^+} W_{i,j} + \sum_{s_i,\, s_j \in V^-} W_{i,j}\right) - C_2 + C_1 \qquad \text{(A.5)}$$

where C_2 is given by $C_2 = \sum_{i=0}^{N}\sum_{j=0}^{N} W_{i,j}$ and C_2 is a constant parameter.

Because C_1 and C_2 are constant parameters, minimizing f(s) is equivalent to minimizing the summation of the weights whose vertices belong to the same partition set. Consequently the maximum clique problem in a quadratic zero-one form can be finally formulated by

$$\text{minimize } E = \sum_{x=0}^{N}\sum_{y=0}^{N}\sum_{i=1}^{2} W(x,y)\, V_{x,i}\, V_{y,i}$$

$$\text{subject to} \sum_{i=1}^{2} V_{x,i} = 1 \text{ for } x = 0 \text{ to } N$$

$$V_{x,i} \in \{1, 0\} \text{ for } x = 1 \text{ to N and } V_{0,1} = 1 \qquad \text{(A.6)}$$

Note that W(x,y) represents the weight of an edge between the xth vertex and the yth vertex, and W(x,x)=0. If $V_{x,i} = 1$, the xth vertex belongs to the ith partition.

Appendix B:

Theorem 1 and theorem 2 are introduced to prove that the proposed system is always allowed to converge to the stable status.

Theorem 1:

$\frac{dE}{dt} \leq 0$ is satisfied under two conditions such as $\frac{dU_i}{dt} = -\frac{\partial E}{\partial V_i}$ and $V_i = f(U_i)$ where $f(U_i)$ is a nondecreasing function.

Proof :

$$\frac{dE}{dt} = \sum_i \frac{dU_i}{dt}\frac{dV_i}{dU_i}\frac{\partial E}{\partial V_i}$$

$$= -\sum_i \left(\frac{dU_i}{dt}\right)^2 \frac{dV_i}{dU_i} \quad \text{where } \frac{\partial E}{\partial V_i} \text{ is replaced by } -\frac{dU_i}{dt} \text{ (condition 1)}$$

$$\leq 0 \qquad \text{where } \frac{dV_i}{dU_i} > 0 \text{ (condition 2)}$$ **Q.E.D.**

In theorem 2, the convergence of the maximum neural network for the maximum clique problem is given.

Theorem 2:

$\frac{dE}{dt} \leq 0$ is satisfied under two conditions such as

(1) $\frac{dU_{x,i}}{dt} = -\frac{\partial E}{\partial V_{x,i}}$ and

(2) $V_{x,i} = 1$ if $U_{x,i} = \max\{U_{x,1}, U_{x,2}\}$
0 otherwise

Proof:

Consider the derivatives of the computational energy E with respect to time t.

$$\frac{dE}{dt} = \sum_x \sum_i \frac{dU_{x,i}}{dt} \frac{dV_{x,i}}{dU_{x,i}} \frac{\partial E}{\partial V_{x,i}}$$

$$= -\sum_x \sum_i \left(\frac{dU_{x,i}}{dt}\right)^2 \frac{dV_{x,i}}{dU_{x,i}}$$ where $\frac{\partial E}{\partial V_{x,i}}$ is replaced

by $-\frac{dU_{x,i}}{dt}$ (condition 1)

Let $\frac{dU_{x,i}}{dt}$ be $\frac{U_{x,i}(t+dt) - U_{x,i}(t)}{dt}$. Let $\frac{dV_{x,i}}{dU_{x,i}}$ be $\frac{V_{x,i}(t+dt) - V_{x,i}(t)}{U_{x,i}(t+dt) - U_{x,i}(t)}$. Let us consider the term $\sum_i \left(\frac{dU_{x,i}}{dt}\right)^2 \frac{dV_{x,i}}{dU_{x,i}}$ for each module separately. Let $U_{x,a}(t+dt)$ be the maximum at time t+dt and $U_{x,b}(t)$ be the maximum at time t for the vertex x.

$$U_{x,a}(t+dt) = \max\{U_{x,1}(t+dt), U_{x,2}(t+dt)\}$$

$$U_{x,b}(t) = \max\{U_{x,1}(t), U_{x,2}(t)\}$$

It is necessary and sufficient to consider the following two cases :

1) $a = b$

2) $a \neq b$

If the condition 1) is satisfied, then there is no state change for the vertex x. Consequently, $\sum_i \left(\frac{dU_{x,i}}{dt}\right)^2 \frac{dV_{x,i}}{dU_{x,i}}$ must be zero.

If 2) is satisfied, then

$$\sum_i \left(\frac{dU_{x,i}}{dt}\right)^2 \frac{dV_{x,i}}{dU_{x,i}} = \left(\frac{U_{x,a}(t+dt)-U_{x,a}(t)}{dt}\right)^2 \frac{V_{x,a}(t+dt)-V_{x,a}(t)}{U_{x,a}(t+dt)-U_{x,a}(t)}$$

$$+\left(\frac{U_{x,b}(t+dt)-U_{x,b}(t)}{dt}\right)^2 \frac{V_{x,b}(t+dt)-V_{x,b}(t)}{U_{x,b}(t+dt)-U_{x,b}(t)}$$

$$= \left(\frac{U_{x,a}(t+dt)-U_{x,a}(t)}{dt}\right)^2 \frac{1}{U_{x,a}(t+dt)-U_{x,a}(t)}$$

$$+\left(\frac{U_{x,b}(t+dt)-U_{x,b}(t)}{dt}\right)^2 \frac{-1}{U_{x,b}(t+dt)-U_{x,b}(t)}$$

$$= \frac{U_{x,a}(t+dt) - U_{x,a}(t)}{(dt)^2} - \frac{U_{x,b}(t+dt) - U_{x,b}(t)}{(dt)^2}$$

$$= \frac{1}{(dt)^2}\{ U_{x,a}(t+dt) - U_{x,a}(t) - U_{x,b}(t+dt) + U_{x,b}(t) \}$$

$$= \frac{1}{(dt)^2}\{ U_{x,a}(t+dt) - U_{x,b}(t+dt) + U_{x,b}(t) - U_{x,a}(t) \}$$

$$> 0$$

because $U_{x,a}(t+dt)$ is the maximum at time t+dt and $U_{x,b}(t)$ is the maximum at time t for the vertex x.

The contribution from each term is either 0 or positive, therefore

$$\sum_i \left(\frac{dU_{x,i}}{dt}\right)^2 \frac{dV_{x,i}}{dU_{x,i}} \geq 0 \text{ and } -\sum_x \sum_i \left(\frac{dU_{x,i}}{dt}\right)^2 \frac{dV_{x,i}}{dU_{x,i}} \leq 0 \quad \Rightarrow \quad \frac{dE}{dt} \leq 0$$

Q.E.D.

To use the maximum neural network for solving the maximum clique problem, the energy function in Eq.(8) and the motion equation in Eq.(9) are used. The convergence property of the maximum neural network is guaranteed by the theorem 1 and theorem 2. The following theorem presents the relationship between the set of local/global minima and the maximal/maximum clique in a given graph.

Theorem 3:

Given a graph G, a graph G_M is defined by following Eq.(A.3) and Eq.(A.4). By minimizing the weight summation over the same partition in a newly constructed graph. G_M, every local/global minimum corresponds to a maximal/maximum clique in the graph G if the maximum neural network is used.

Proof:

Suppose that the system reaches an equilibrium state, i.e. local/global minimum, at t=T. Then there exists a partition of V^+ and V^- in the graph G_M. Because the system stays in the equilibrium state and the maximum neural network is used, the following conditions must be satisfied

$$\text{for } i \in V^+,\ V_{i,1}(T) = 1 \text{ and } \frac{dU_{i,1}(T)}{dt} > \frac{dU_{i,2}(T)}{dt}$$

$$\text{for } i \in V^-,\ V_{i,2}(T) = 1 \text{ and } \frac{dU_{i,1}(T)}{dt} < \frac{dU_{i,2}(T)}{dt}$$

It is defined that if there exists an edge between the *i*th vertex and the *j*th vertex then (i,j)=1 else (i,j)=0.

Suppose that the final state does not correspond to a clique in the graph G, then

$\exists$ vertex a and vertex b $\in V^+$, and there exists an edge between the *a*th vertex and the *b*th vertex in $\overline{G}$ shown in Fig. A.1.

Let $S=\{i \mid i \in V^+ \text{ and } (i,a)=1 \text{ in } \overline{G}\}$ and $Q=\{i \mid i \in V^- \text{ and } (i,a)=1 \text{ in } \overline{G}\}$.

$\Rightarrow |S| \geq 1$ where $|S|$ denotes the number of elements in S.

Because $a \in V^+$

$$\Rightarrow \frac{dU_{a,1}(T)}{dt} > \frac{dU_{a,2}(T)}{dt}$$

$$\Rightarrow \sum_{y=0}^{N} W_{a,y} V_{y,1} < \sum_{y=0}^{N} W_{a,y} V_{y,2} \quad \text{where } W_{i,j} \text{ is defined in Eq.(A.4)}$$

$$\Rightarrow W_{a,0} + \sum_{y=1}^{N} W_{a,y} V_{y,1} < \sum_{y=1}^{N} W_{a,y} V_{y,2} \quad (\text{because } s_0 \in V^+)$$

$$\Rightarrow \frac{1}{4}(\deg(a) + c_a) + \frac{1}{4}|S| < \frac{1}{4}|Q|$$ where $\deg(a)$ denotes the degree of the *a*th vertex in the graph $\overline{G}$.

$$\Rightarrow \frac{1}{4}(|Q| + |S| - 1) + \frac{1}{4}|S| < \frac{1}{4}|Q|$$ where c_a is equal to -1.

$$\Rightarrow |S| < \frac{1}{2}$$

However, $|S| \geq 1$ then we reach a contradiction.

Therefore, every local/global minimum for the max cut in the graph G_M corresponds to a maximal/maximum clique in the graph G if the maximum neural network is used for solving the maximum clique problem. The proof is complete. **Q.E.D.**

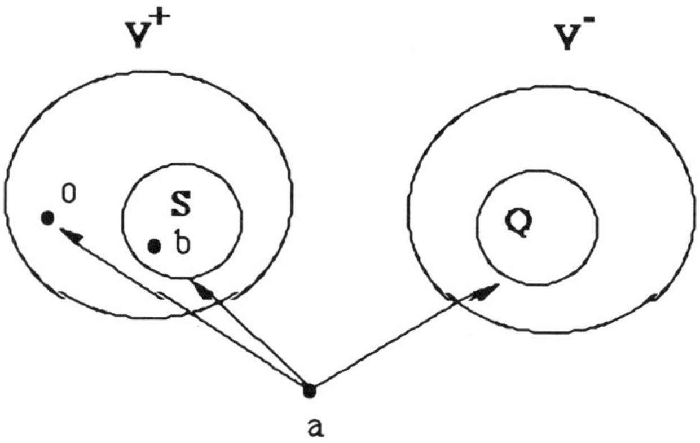

Fig. A.1 An illustration for the Theorem 3.

Appendix C:

Solution quality comparisons between our algorithm and Carraghan and Pardalos' algorithm where each element printed on boldface type represents the maximum clique size found by Carraghan and Pardalos' algorithm and that on lightface type represents the solution found by our algorithm. The 'x' represents the running time is not applicable because it takes prohibitively long time.

	Density								
Size	**0.1**	**0.2**	**0.3**	**0.4**	**0.5**	**0.6**	**0.7**	**0.8**	**0.9**
10	**2**	**2**	**3**	**3**	**4**	**5**	**5**	**5**	**6**
	2	2	3	4	5	6	6	6	9
20	**2**	**3**	**4**	**5**	**6**	**7**	**7**	**9**	**11**
	2	3	4	5	6	7	8	9	13
30	**3**	**3**	**5**	**5**	**6**	**7**	**10**	**10**	**15**
	3	3	5	6	7	8	10	11	16
40	**3**	**4**	**5**	**6**	**7**	**8**	**10**	**13**	**17**
	3	4	5	6	8	8	11	14	19
50	**3**	**4**	**5**	**6**	**7**	**8**	**12**	**14**	**20**
	3	4	5	6	7	9	11	13	21
60	**3**	**4**	**5**	**7**	**8**	**10**	**11**	**15**	**20**
	3	4	5	6	8	10	13	15	24
70	**4**	**4**	**6**	**7**	**8**	**11**	**12**	**17**	**25**
	4	4	6	7	8	10	14	18	23
80	**4**	**4**	**5**	**7**	**8**	**11**	**14**	**17**	**24**
	4	4	5	7	9	11	13	18	28
90	**4**	**5**	**6**	**7**	**9**	**11**	**14**	**18**	**27**
	4	4	5	7	9	10	13	18	25
100	**4**	**5**	**6**	**7**	**9**	**11**	**14**	**21**	**28**
	3	5	6	7	9	11	13	18	28
110	**4**	**5**	**7**	**7**	**9**	**12**	**15**	**20**	**30**
	4	5	7	7	8	11	14	20	30
120	**5**	**5**	**6**	**8**	**10**	**12**	**15**	**20**	**32**
	4	5	6	8	9	11	14	20	30
130	**4**	**5**	**6**	**8**	**10**	**12**	**17**	**21**	**30**
	4	5	5	7	9	11	14	21	31
140	**4**	**6**	**7**	**8**	**10**	**13**	**16**	**22**	**x**
	4	5	6	7	9	11	14	20	32
150	**4**	**6**	**7**	**8**	**10**	**12**	**17**	**23**	**x**
	4	5	7	8	9	11	15	19	33
160	**4**	**6**	**7**	**8**	**11**	**13**	**17**	**22**	**x**
	4	5	6	8	11	12	16	20	33
170	**4**	**6**	**7**	**8**	**11**	**13**	**16**	**23**	**x**
	4	5	6	8	9	12	15	21	34
180	**4**	**5**	**7**	**8**	**11**	**13**	**17**	**x**	**x**
	4	5	7	8	10	11	16	21	35
190	**4**	**6**	**7**	**9**	**11**	**14**	**17**	**x**	**x**
	4	5	6	7	9	12	16	21	33
200	**4**	**6**	**7**	**9**	**10**	**14**	**18**	**x**	**x**
	4	5	7	7	9	11	15	21	36
210	**4**	**6**	**7**	**9**	**11**	**14**	**18**	**x**	**x**
	4	5	7	8	11	12	15	21	35
220	**5**	**6**	**7**	**9**	**11**	**14**	**18**	**x**	**x**
	4	6	7	8	10	12	16	21	37
230	**4**	**6**	**7**	**9**	**11**	**14**	**18**	**x**	**x**
	4	5	7	8	10	13	17	21	36
240	**4**	**6**	**7**	**9**	**11**	**14**	**19**	**x**	**x**
	4	5	6	8	10	12	15	22	37
250	**4**	**6**	**7**	**9**	**11**	**14**	**18**	**x**	**x**
	4	5	7	8	10	14	16	23	39
260	**5**	**6**	**8**	**9**	**11**	**14**	**19**	**x**	**x**
	4	5	7	8	10	12	16	24	38
270	**5**	**6**	**8**	**9**	**12**	**15**	**19**	**x**	**x**
	4	5	7	8	10	13	16	23	41
280	**4**	**6**	**8**	**10**	**12**	**15**	**20**	**x**	**x**
	4	6	7	8	10	13	17	23	40
290	**4**	**6**	**8**	**9**	**12**	**15**	**20**	**x**	**x**
	4	5	7	8	11	13	17	23	40
300	**5**	**6**	**8**	**10**	**12**	**15**	**19**	**x**	**x**
	4	6	7	9	10	12	17	23	39

Chapter 4

MAXIMUM CLIQUE PROBLEMS: PART 2

NOBUO FUNABIKI
Sumitomo Metal Ind.
Osaka, Japan

YOSHIYASU TAKEFUJI
Dept. of Electrical Engineering and Applied Physics
Case Western Reserve University
Cleveland, OH 44106, USA
and
Faculty of Environmental Information
Keio University, Fujisawa 252, Japan

A parallel algorithm based on the neural network model for finding a near-maximum clique is presented in this chapter. A maximum clique of a graph G is a maximum complete subgraph of G, where any two vertices are adjacent. The problem of finding a maximum clique is NP-complete. The parallel algorithm requires n processing elements for an n-vertex graph problem. The algorithm is verified by solving 230 different graph problems. The simulation results show that our computation time on a Macintosh IIfx is shorter than that of two best known algorithms on Cray 2 and IBM 3090, while the solution quality is similar. The algorithm solves a near-maximum clique problem in nearly constant time on a parallel machine with n processors. This chapter is based on the paper published in Journal of Parallel and Distributed Computing (Funabiki et al. 1992).

4.1 INTRODUCTION

A parallel algorithm based on the neural network model for finding a near-maximum clique of an arbitrary graph is presented. When a graph G with a set of vertices and a set of edges has an edge between two vertices, they are called adjacent. A clique of G is a complete subgraph of G, where any two vertices are adjacent to each other. The maximum clique is a clique with the largest number of vertices among cliques of G. The problem of finding a maximum clique of an arbitrary graph is known to be NP-complete (Garey M. R. and Johnson D. S. 1979). This means that in the worst case the computing time for solving the maximum clique problem grows exponentially with the graph size. Fig. 1 (a) shows an 8-vertex-17-edge graph (Carraghan R. and Pardalos P. M. 1990). The maximum cliques are {2, 3, 4, 7} and {2, 4, 6, 7} as shown in Fig. 1 (b) and (c).

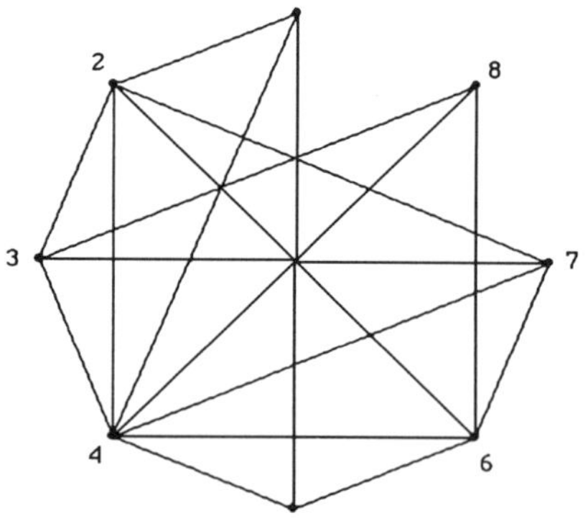

Fig. 1 (a) An 8-vertex-17-edge graph problem.

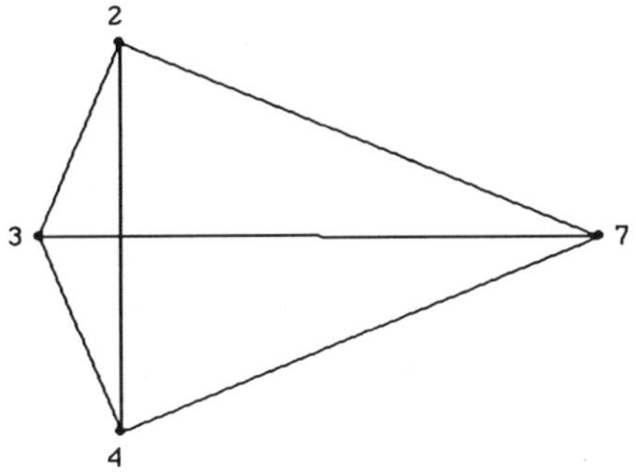

Fig. 1 (b) The maximum clique #1.

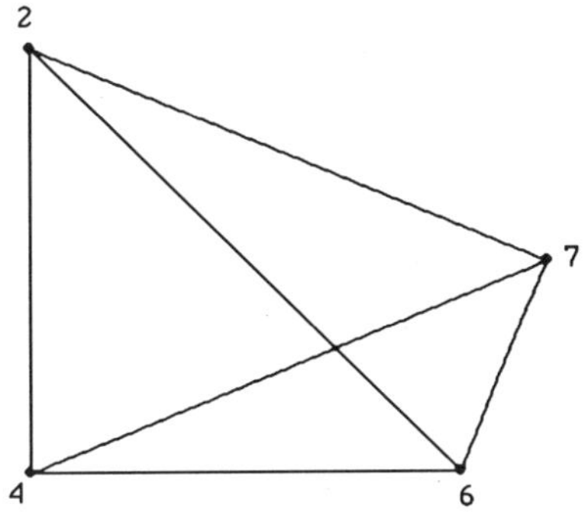

Fig. 1 (c) The maximum clique #2.

Several sequential algorithms for the maximum clique problem have been reported, where the computation time depends not only on the number of vertices in the graph, but also on the number of edges. They can deal neither with large size graphs nor with high density graphs. Note that the density means the ratio between the number of edges in the n-vertex graph and that in the n-vertex complete graph n(n-1)/2. In 1986, Balas and Yu proposed an O(n+m) time branch and bound algorithm for an n-vertex-m-edge graph, where it was only tested by small size graphs with maximally 400 vertices and 30,000 edges (Balas E. and Yu C. S. 1986). In 1990, Pardalos and Phillips formulated the maximum clique problem as a linearly constrained indefinite quadratic global optimization problem (Pardalos P. M. and Phillips A. T. 1990). Although the supercomputer Cray 2 was used, their algorithm could not solve larger than 75-vertex graph problems. It required more than one hour even for the 75-vertex-91%-edge-density graph problem on Cray 2. In 1990, Carraghan and Pardalos proposed an algorithm based on a partial enumeration (Carraghan R. and Pardalos P. M. 1990). Although it could solve up to 3000-vertex and over 1,000,000-edge graph problems on the main frame IBM 3090, it required prohibitively long computation time even for a middle size graph problem. For example, it took more than one hour for a 1000-vertex-40%-density graph problem on IBM 3090. In 1980, Mead and Conway proposed a parallel algorithm for the maximum clique problem, where it requires $O(n^2)$ time with maximum $O(2^n)$ processors for an n-vertex graph problem (Mead C. and Conway L. 1980).

Several applications of the maximum clique problem for real world problems have been reported. Balas and Yu introduced four applications: information retrieval, experimental design, signal transmission, and computer vision (Balas E. and Yu C. S. 1986). Ogawa proposed an application for the labeled point pattern matching problem (Ogawa H. 1986). Lecky et al. proposed an application for the PLA folding problem (Lecky J. E. et al. 1989). Horaud and Skordas proposed an application for the stereo vision correspondence problem (Horaud R. and Skordas T. 1989).

While existing algorithms require long computation time on expensive machines, our algorithm can solve large size problems both on an inexpensive machine such as a personal computer and on a parallel machine. It can also be realized on an analog circuit (Takefuji Y. and Lee K. C. 1991). The computation time and the number of required processing elements in our algorithm do not depend on the graph density.

4.2 NEURAL NETWORK APPROACH

The neural network model is composed of n processing elements (neurons) for an n-vertex graph problem. Processing element #i has an input U_i and an output V_i. The

McCulloch-Pitts neuron model (McCulloch W. S. and Pitts W. H. 1943) is adopted in this chapter, where the input/output function is given by:

$$V_i=1 \text{ if } U_i>0, \quad V_i=0 \text{ otherwise} \tag{1}$$

The change of U_i is given by the motion equation in order to minimize the energy function $E(V_1,\ldots,V_n)$, which is determined by considering the necessary and sufficient constraints in the problem (Funabiki N. and Takefuji Y. 1991) (Hopfield J. J. and Tank D. W. 1985) (Takefuji Y. and Lee K. C. 1989) (Takefuji Y. and Lee K. C. 1990) (Takefuji Y. and Lee K. C. 1991). The motion equation is given by:

$$\frac{dU_i}{dt} = -\frac{\partial E(V_1,\ldots,V_n)}{\partial V_i} \tag{2}$$

It has been proven that the motion equation forces the neural network system to converge to the local minimum (Takefuji Y. and Lee K. C. 1991).

Fig. 2 shows the neural network representation for the maximum clique problem in Fig. 1 (a), where a total of eight processing elements are required. The output of processing element #i represents whether vertex #i belongs to the clique (solution) or not. The nonzero output (V_i=1) means that vertex #i belongs to the clique, while the zero output (V_i=0) means not. Fig. 2 shows the solution state corresponding to Fig. 1 (b), where the black (white) square indicates the nonzero (zero) output.

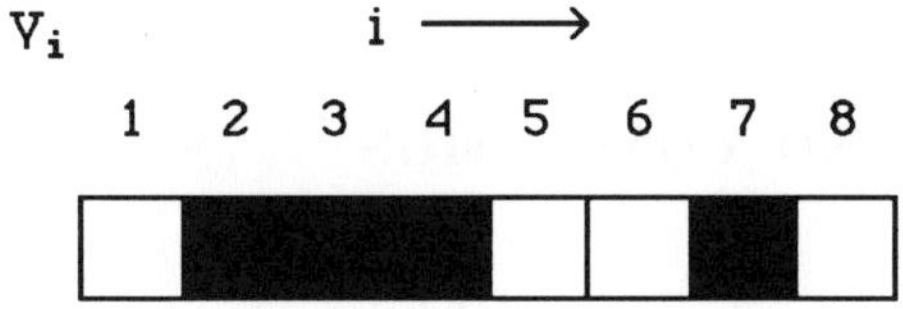

Fig. 2 Neural network representation for the maximum clique problem in Fig. 1.

Our approach to the maximum clique problem is to maximize the number of vertices of a selected complete subgraph. The output of processing element #i be zero if vertex #i is not adjacent to a vertex in a complete subgraph, and the output be nonzero if vertex #i is adjacent to all vertices in the subgraph. The motion equation for processing element #i in the n-vertex graph problem is composed of two terms: a negative force and a positive force, which is given by:

$$\frac{dU_i}{dt}=-A\sum_{j=1}^{n}(1-d_{ij})V_j+Bh\left(\sum_{j=1}^{n}(1-d_{ij})V_j+V_i\right) \quad (3)$$

where d_{ij} is 1 if two vertices #i and #j are adjacent to each other, 0 otherwise. Note that $d_{ij}=d_{ji}$ and $d_{ii}=1$ are always satisfied. The A-term discourages processing element #i from having the nonzero output if vertex #i is not adjacent to a vertex chosen in a clique. The B-term encourages processing element #i to have the nonzero output if vertex #i is adjacent to all vertices in the clique and the output of processing element #i is zero. The function h(x) is 1 if x=0, 0 otherwise. A and B are constant coefficients. Because a vertex, which has many adjacent vertices, is more likely to belong to the maximum clique, the coefficient B is changed by the number of adjacent vertices in our algorithm:

$$B=\frac{\text{the number of adjacent vertices}}{n\times(\text{density of the graph})}\times 20 \quad (4)$$

A=1 is used in our simulation.

4.3 SIMULATION AND CONCLUSION

The following procedure describes the parallel algorithm for finding a near-maximum clique of an n-vertex graph based on the first order Euler method.

0. Set t=0.

1. The initial values of $U_i(t)$ for i=1,...,n are uniformly randomized between 0 and -20, and the initial values of $V_i(t)$ for i=1,...,n are assigned to 0.

2. $$\Delta U_i(t) = -A \sum_{j=1}^{n} (1-d_{ij}) V_j(t) + Bh\left(\sum_{j=1}^{n} (1-d_{ij}) V_j(t) + V_i(t) \right) \tag{5}$$

3. $$U_i(t+1) = U_i(t) + \Delta U_i(t) \tag{6}$$

4. $V_i(t+1)=1$ if $U_i(t+1)>0$, $V_i(t+1)=0$ otherwise (7)

5. If $\left(V_i(t)=1 \text{ and } \sum_{j=1}^{n} (1-d_{ij}) V_j(t) = 0 \right)$ or $\left(V_i(t)=0 \text{ and } \sum_{j=1}^{n} (1-d_{ij}) V_j(t) \neq 0 \right)$
for i=1,...,n then terminate this procedure, else increment t by 1 and go to step 2.

The simulator has been developed on Macintosh IIfx in order to verify our algorithm. 4 graph size problems (10, 25, 50, and 75 vertices) with 5 edge densities (10%, 25%, 50%, 75%, and 90%) were simulated to compare our algorithm with Parados's algorithm on Cray 2 (Pardalos P. M. and Phillips A. T. 1990), where 10 different graphs were randomly generated for each size-density and 10 simulation runs were performed from different initial values of $U_i(t)$ for each graph problem. Table 1-4 show the results on the solution quality and the computation time (seconds). Our computation time on Macintosh IIfx is shorter than theirs on Cray 2, while the solution quality is about the same. Note that the speed of Cray 2 is 125 MIPS, while that of Macintosh IIfx is 7 MIPS.

Randomly generated 10 graph size problems (100-1000 vertices) with 3 edge densities (25%, 50%, and 75%) were simulated to compare our algorithm with Carraghan's algorithm on IBM 3090 (Carraghan R. and Pardalos P. M. 1990) where

100 simulation runs were performed from different initial values of $U_i(t)$ for each problem. Fig. 3 and 4 show respectively 10-vertex and 25-vertex graph problems and our maximum clique solutions.

Table 1 Comparisons of simulation results for 10-vertex problems.

Pardalos & Phillips's algorithm on Cray 2			Our algorithm on Macintosh IIfx		
average density	average clique size	average comp. time	average density	average clique size	average comp. time
0.10	2.1	0.3	0.06	2.0	0.03
0.22	2.9	0.5	0.20	2.5	0.03
0.50	3.7	0.6	0.53	3.9	0.03
0.75	5.5	0.4	0.76	5.4	0.03
0.88	7.0	0.2	0.93	7.5	0.03

Table 2 Comparisons of simulation results for 25-vertex problems.

Pardalos & Phillips's algorithm on Cray 2			Our algorithm on Macintosh IIfx		
average density	average clique size	average comp. time	average density	average clique size	average comp. time
0.09	2.6	6.7	0.09	2.4	0.04
0.24	3.4	7.3	0.25	3.3	0.04
0.49	5.5	7.3	0.53	5.3	0.05
0.75	8.7	6.7	0.78	9.3	0.06
0.91	14.2	4.7	0.91	13.8	0.06

Table 3 Comparisons of simulation results for 50-vertex problems.

Pardalos & Phillips's algorithm on Cray 2			Our algorithm on Macintosh IIfx		
average density	average clique size	average comp. time	average density	average clique size	average comp. time
0.10	2.7	93.6	0.09	2.7	0.12
0.25	3.6	98.1	0.25	3.9	0.13
0.50	5.9	86.3	0.51	6.7	0.13
0.75	10.4	78.7	0.77	11.7	0.15
0.90	18.8	69.2	0.89	18.9	0.16

Table 4 Comparisons of simulation results for 75-vertex problems.

Pardalos & Phillips's algorithm on Cray 2			Our algorithm on Macintosh IIfx		
average density	average clique size	average comp. time	average density	average clique size	average comp. time
0.09	3.0	2283.0	0.10	2.8	0.24
0.24	4.0	3690.1	0.25	4.0	0.24
0.49	6.0	2837.1	0.51	6.9	0.30
0.75	11.0	2941.4	0.76	13.1	0.30
0.91	22.0	3879.6	0.90	23.0	0.36

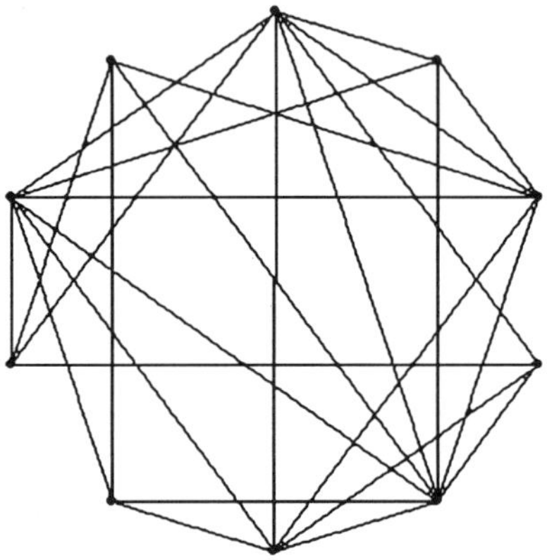

Fig. 3 (a) A 10-vertex graph problem.

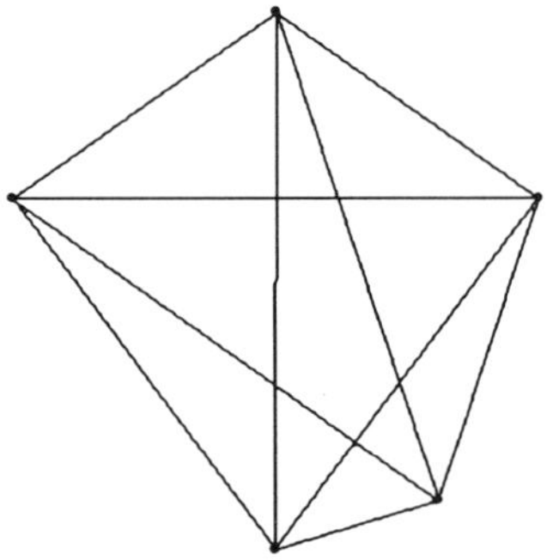

Fig. 3 (b) A maximum clique solution of the graph.

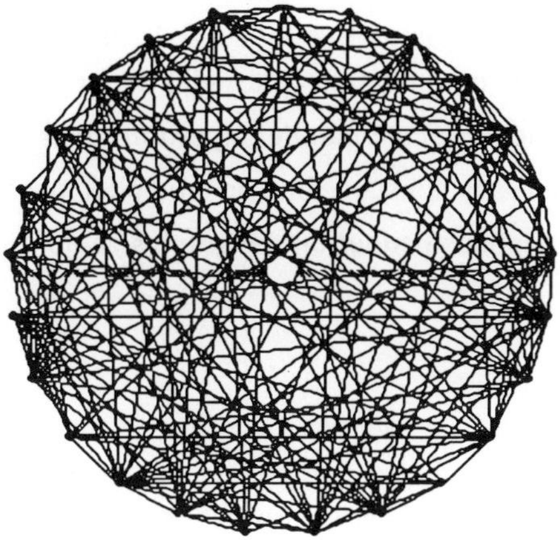

Fig. 4 (a) A 25-vertex graph problem.

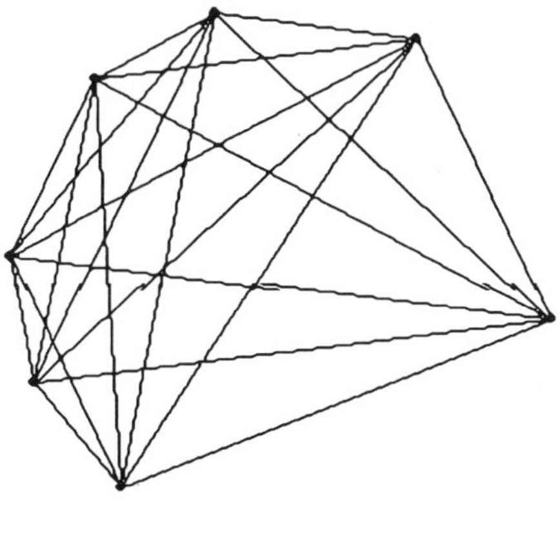

Fig. 4 (b) A maximum clique solution of the graph.

Table 5 shows the average number of iteration steps to converge to solutions, and the average/maximum numbers of vertices in cliques found by our simulator. Table 6 compares the computation time (seconds) between our algorithm on Macintosh IIfx and Carraghan's algorithm on IBM 3090. Fig. 5 shows the distribution of the number of iteration steps required to converge to solutions in the 1000-vertex-50%-density-edge problem.

Table 1-4 and 6 show that our algorithm is superior to the best existing algorithms in terms of the computation time with about the same solution quality. We also conclude that the presented parallel algorithm finds a near-maximum clique of an n-vertex graph in nearly constant time on a parallel machine with n processors. The computation time depends neither on the graph size nor on the graph density on a parallel machine.

Table 5 Summary of simulation results for 30 graph problems.

graph size	25% density		50% density		75% density	
	ave. steps	sol. quality	ave. steps	sol. quality	ave. steps	sol. quality
100	16.0	4.2 / 5	15.8	8.0 / 9	19.3	14.1 / 16
200	18.8	4.9 / 7	23.9	8.5 / 10	23.4	15.9 / 18
300	28.9	5.1 / 6	19.1	8.9 / 11	26.8	15.9 / 18
400	25.7	4.9 / 6	25.7	8.9 / 11	26.8	17.7 / 20
500	26.7	6.2 / 7	33.3	9.4 / 11	31.3	18.9 / 22
600	33.8	5.7 / 7	29.4	9.9 / 12	30.4	19.5 / 23
700	39.8	5.4 / 6	32.4	9.9 / 12	31.3	20.1 / 23
800	41.1	6.0 / 7	38.2	9.9 / 12	31.3	20.5 / 24
900	47.9	5.8 / 7	38.8	10.2 / 12	34.6	21.0 / 24
1000	49.9	5.8 / 7	45.9	10.4 / 12	34.4	21.4 / 25

Table 6 Comparisons of the computation time (seconds) for 50%-edge-density graph problems.

graph size	Carraghan's algorithm on IBM 3090	Our algorithm on Macintosh IIfx
100	0.14	0.52
200	4.16	2.7
300	46.04	4.8
400	235.68	11.2
500	1114.78	22.3
600	-	28.1
700	-	45.0
800	-	69.9
900	-	88.8
1000	-	129.8

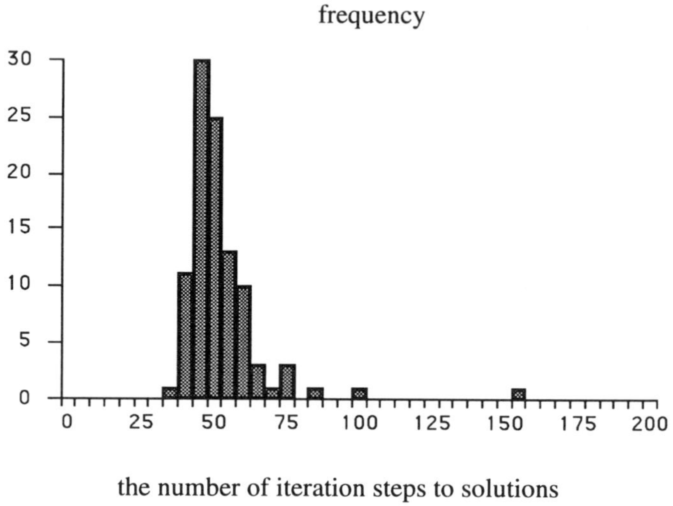

Fig. 5 The distribution of the number of iteration steps required to converge to solutions in the 1000-vertex-50%-density-edge problem.

4.4 REFERENCES

Balas E. and Yu C. S., (Nov. 1986), Finding a maximum clique in an arbitrary graph, SIAM J. Comput., Vol. 15, 4, pp. 1054-1068.

Carraghan R. and Pardalos P. M., (Nov. 1990), An exact algorithm for the maximum clique problem, Operations Research Letters, Vol. 9, pp. 375-382.

Funabiki N. and Takefuji Y., (August 1991), A parallel algorithm for spare allocation problems, IEEE Trans. Reliability, Vol. 40, 3, pp. 338-346.

Funabiki N., Takefuji Y., and Lee K. C., (March 1992), A neural network model for finding a near-maximum clique, Journal of Parallel and Distributed Computing, Vol. 14, 3, pp. 340-344.

Garey M. R. and Johnson D. S., (1979), Computers and Intractability: A Guide to the Theory of NP-Completeness, (W. H. Freeman, New York).

Hopfield J. J. and Tank D. W., (1985), Neural computation of decisions in optimization problems, Biological Cybernetics, Vol. 52, pp. 141-152.

Horaud R. and Skordas T., (Nov. 1989), Stereo correspondence through feature grouping and maximal cliques, IEEE Trans. Pattern Analy. Mach. Intelligence, Vol. 11, 11, pp. 1168-1180.

Lecky J. E., Murphy O. J., and Absher R. G., (Sept. 1989), Graph theoretic algorithms for the PLA folding problem, IEEE Trans. CAD/ICAS, Vol. 8, 9, pp. 1014-1021.

McCulloch W. S. and Pitts W. H., (1943), A logical calculus of ideas immanent in nervous activity, Bulletin of Mathematical Biophysics, Vol. 5, pp. 115-133.

Mead C. and Conway L., (1980), Introduction to VLSI systems, (Addison-Wesley).

Ogawa H., (1986), Labeled point pattern matching by Delaunay triangulation and maximal cliques, Pattern Recognition, Vol. 19, 1, pp. 35-40.

Pardalos P. M. and Phillips A. T., (1990), A global optimization approach for solving the maximum clique problem, Int. J. Computer Mathematics, Vol. 33, pp. 209-216.

Takefuji Y. and Lee K. C., (Sept. 1989), A near-optimum parallel planarization algorithm, Science, 245, pp. 1221-1223.

Takefuji Y. and Lee K. C., (March 1990), A parallel algorithm for tiling problems, IEEE Trans. Neural Networks, Vol. 1, 1, pp. 143-145.

Takefuji Y. and Lee K. C., (March 1991), Artificial neural networks for four-coloring map problems and K-colorability problems, IEEE Trans. Circuits Systems, Vol. 38, 3, pp. 326-333.

Chapter 5

MULTI-LAYER CHANNEL ROUTING PROBLEMS

KYOTARO SUZUKI
HIDEHARU AMANO
YOSHIYASU TAKEFUJI
Keio University
Japan

Multi-layer channel routing is one of cumbersome jobs in automatic layout design of VLSI chips and PCBs. As VLSI chips have been used in every field of electrical engineering, it becomes more important to reduce the layout design time. With the advancement of the VLSI technology, four-layer problems can be treated and the algorithms for more than four-layer problems will be demanded in the near future. Proposed algorithm can treat $2 \times n$-layer problems in parallel. In the paper, this algorithm is introduced and implemented on a multiprocessor system. By minimizing the communication overhead and load unbalance between processors, the performance with 8 processors is improved by between 6 and 6.5 times compared with the sequential version.

5.1 INTRODUCTION

Multi-layer channel routing problems are proved to be NP-complete and one of cumbersome jobs in automatic layout design of VLSI chips and PCBs. The problem is not only to route the given interconnections between terminals in two parallel horizontal rows on the multi-layer channel without overlapping each other, but also to minimize the channel area. Many sequential algorithms for the two-layer channel routing problems have been proposed (Hashimoto A. and Stevens J. 1971) (Kernighan B. W. and Schweikert D. G. 1973) (Deutsch D. N. 1976) (Sahni S. and Bhatt A. 1980) (LaPaugh A. S. 1980) (Dolev D. et al. 1981) (Yoshimura T. and Kuh E. S. 1982) (Rivest R. L. and Fiduccia C. M. 1982) (Leong H. W. and Liu C. L. 1983) (Burstein M. and Pelavin R. 1983) (Szymanski T. G. 1985) (Reed J., Sangiovanni-Vincentelli A., and Santomauro M. 1985) (Leong H. W., Wong D. F., and Liu C. L. 1985) (Joobbani R. and Siewiorek D. P. 1985). Each set of terminals must be interconnected through a routing path which consists of only a horizontal segment and several vertical segments. The horizontal segments and the vertical segments must be routed on different layers respectively. In some algorithms (Deutsch D. N. 1976) (Yoshimura T. and Kuh E. S. 1982) (Rivest R. L. and Fiduccia C. M. 1982) (Leong H. W. and Liu C. L. 1983) (Burstein M. and Pelavin R. 1983) (Szymanski T. G. 1985) (Reed J., Sangiovanni-Vincentelli A., and Santomauro M. 1985) (Leong H. W., Wong D. F., and Liu C. L. 1985) (Joobbani R. and Siewiorek D. P. 1985), doglegging has been introduced, where only a horizontal segment of a routing path is split into more than two horizontal segments. into more than two horizontal segments in order to minimize the channel area. In general, doglegging increases the number of via holes in spite of reducing of the channel area. Doglegging reduces the reliability of the VLSI system and increases the manufacturing cost. Cyclic conflict occurs when net A has a top terminal at the same column where net B has a bottom terminal while net A has a bottom terminal at the same column where net B has the top terminal (Fig. 1). For the problems with the cyclic conflict, routing paths cannot be routed without doglegging or more than two layers. The cyclic conflict occurs infrequently and most of them can be avoided by rearranging the terminal placement.

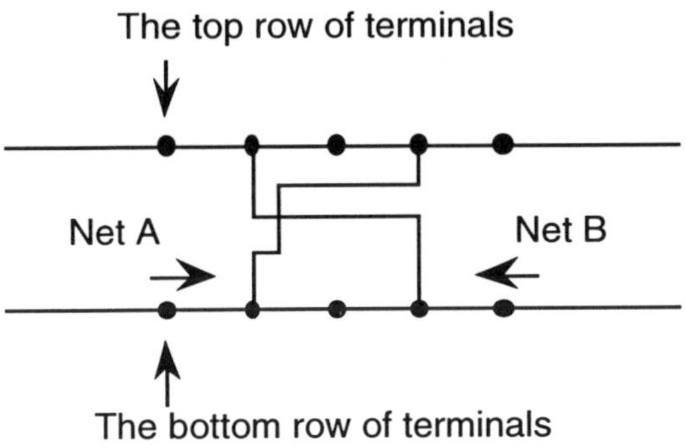

Fig. 1 Cyclic problem.

In order to reduce the channel area further, sequential algorithms for two-layer-and-over-the-cell channel routing problems (Deutsch D. N and P. Glick 1980) (Krohn H. E. 1983) (Shiraishi Y. and Sakemi J. 1987) (Gudmundsson G. and Ntafos S. 1987) (Cong J., Wong D. F., and Liu C. L. 1988) (Cong J. and Liu C. L. 1990), and three-layer channel routing problems (Chen Y. K. and M. L. Liu 1984) (Cong J., Wong D. F., and Liu C. L. 1987) have been proposed. The over-the-cell algorithms use not only the channel area but also the area over the cells for interconnections. Although these algorithms can find better solutions, they are still based on two-layer routing technique, so that it is not easy to apply them for more than three-layers problems.

As VLSI chips have been used in every field of electrical engineering, it becomes more important to reduce the layout design time. Several parallel routing algorithms for channel routing problems and two-dimensional routing algorithms have been proposed and implemented on parallel processor systems (Chang S. C. and J. aJ. a J. 1988a) (Chang S. C. and J\. aJ\. a J. 1988b) (Chang S. C. and J\. aJ\. a J. 1988c) (Krishnamurthy S. and J\. aJ\. a J. 1988) (Iosupovici A. 1986) (Rose J. 1990) (Brouwer R. J. and Banerjee P. 1988) (Watanabe T., Kitazawa H., and Sugiyama Y. 1987) (Brower R. J. and Banerjee P. 1990) (Takahashi Y. and Sasaki S. 1990).

The algorithm for more than four-layer channel problems is demanded in the near future. Proposed algorithm (Funabiki N. and Takefuji Y. 1992) (Takefuji Y. 1992) can solve more than four-layer problems flexibly and it is suited for parallel computing. The algorithm has been implemented on a multiprocessor system and the performance is evaluated.

5.2 DEFINITIONS

On a chip for channel routing, all standard cells have the same height and stand in rows. A channel consists of two parallel horizontal rows of points which are called terminals and the area between them in Fig. 2.

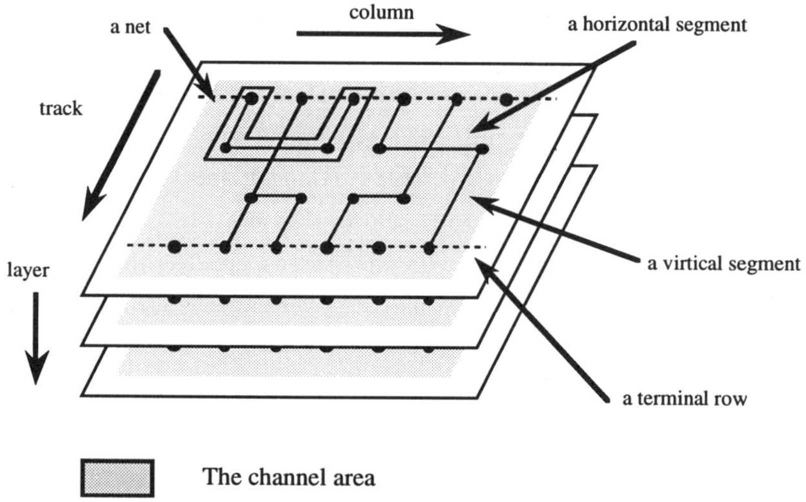

Fig. 2 The channel routing problem

Each terminal is placed at regular intervals and can be distinguished from each other by the column and the end (top end or bottom end) at which the terminal is placed. The coordinate parallel to terminal rows is called the column and the vertical coordinate is called the track.

A net consists of a set of terminals at one or both ends that must be interconnected through some routing paths. A path consists of only a horizontal segment parallel to the terminal rows and several vertical segments. The horizontal segments and the vertical segments must be assigned on different layers respectively. Some pairs of two layers are used for the multi-layer channel routing problem. Each net is routed on a pair of two layers. One is for only a horizontal segment, and the other is for the vertical segments. The connections between those segments are made through via holes. Any two routing paths on the same layer cannot be placed within some distance, which is called separation condition. A unit grid is superimposed on the channel where the size of one unit satisfies the separation condition and all the terminals are located at the grid points. The separation condition is that any two nets must be embedded neither on the same track nor on the same column. If they overlap there, which is called overlapping condition. If the segments of other nets are overlapping on the same grid, the separation condition is not satisfied. All the routing paths on the channel must follow the grid lines. The multi-layer channel routing problem is not only to route the interconnections between the terminals in the same net without overlapping but also to minimize the channel area, that is to minimize the number of tracks as long as the number of layers is constant.

5.3 ALGORITHM

The quality of a multi-layer channel routing algorithm is described by total channel area which is the number of layers multiplied by the number of tracks, the average number of iteration steps to converge on a solution, and convergence frequency for the benchmarks under many conditions. The quality of the proposed algorithm is discussed in (Funabiki N. and Takefuji Y. 1992) (Takefuji Y. 1992).

Fig. 3 shows a channel routing problem in (Yoshimura T. and Kuh E. S. 1982) where ten nets are given to be routed in the four-layer channel which has three tracks on each layer. Each terminal of a net is labeled by the same number. For example, the top terminal at 3rd column and the bottom terminals at 2nd column and at 4th column are in the net 4.

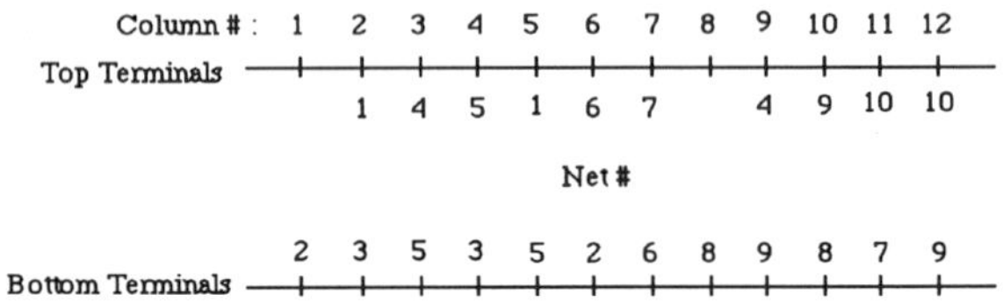

Fig. 3 A 10-net-3-track problem.

The proposed algorithm is based on a three dimensional Hopfield neural network model which consists of a large number of massively interconnected simple neurons. In the algorithm, doglegging and cyclic conflict are not treated because of the reason described in section 1. Only a horizontal segment for a net is embedded, and the vertical segments of the net are automatically assigned on the layer corresponding to the layer on which the horizontal segment of the net is assigned. The channel routing problem can be simplified into the layer-track problem to find the layer number and the track number for embedding the horizontal segment of the given net without violating the overlapping conditions.

Fig. 4 shows the output state of neurons for this problem. $6(=3\times2)$ neurons are used to indicate on which layer and track only a horizontal segment of a given net should be embedded. $60(=10\times3\times2)$ neurons are used in total because 6 neurons are used for each net. Generally, $n\times m\times l$ neurons are used to represent n nets, m tracks, and $2\times l$ layers. L layers are for the horizontal segments and the other l layers are for the vertical segments. V_{ijk} represents the output from the ijk-th neuron which corresponds to the i-th net and the j-th track on the k-th layer for i = 0, .., 10 - 1, j = 0, .., 3 - 1, k = 0, 2 - 1, while U_{ijk} is the input into the ijk-th neuron. The black square ($V_{ijk}=1$) means that the horizontal segment of i-th net should be embedded on the corresponding track and layer while the white square ($V_{ijk}=0$) means opposite. Fig. 5 shows the routing solution corresponding to Fig. 4.

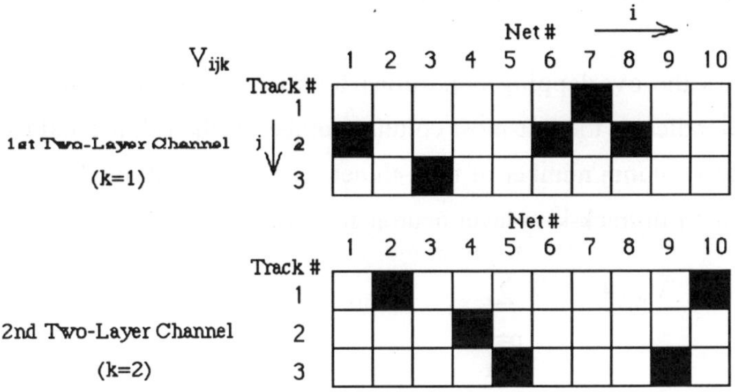

Fig. 4 The output state of neurons.

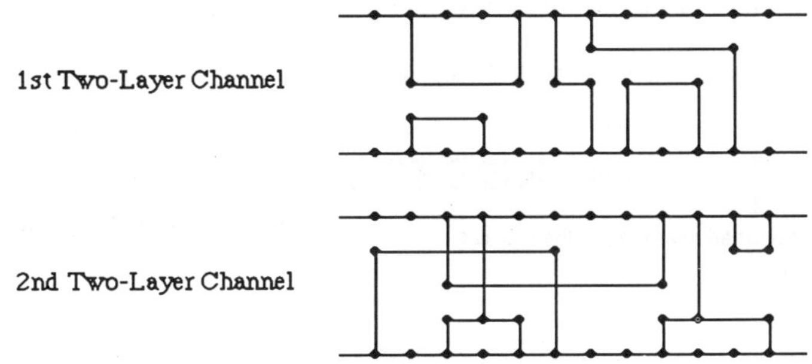

Fig. 5 The routing solution.

DEFINITION OF V_{ijk}

The output V_{ijk} based on the modified McCulloch-Pitts neuron model is described as following:

$V_{ijk}=1$ if $U_{ijk}>0$ and $U_{ijk}=\max\{ U_{iqr} \}$ for q=0,...,m-1 and r=0,...,l-1.

$=0$ otherwise.

THE OVERLAPPING CONDITIONS

Fig. 6 shows the overlapping conditions for the horizontal segments of the nets where head indicates the left-most column number of the i-th net and $tail_i$ indicates the right-most column number of the i-th net. The horizontal overlapping conditions for the i-th-net-j-th-track-k-th-layer neuron are given by:

$$\sum_{\substack{p=0\\ p\neq i\\ head_i\leq head_p\leq tail_i}}^{n-1} V_{pjk} + \sum_{\substack{p=0\\ p\neq i\\ head_p\leq head_i\leq tail_p}}^{n-1} V_{pjk}$$

The condition is nonzero if the horizontal segments of the other nets overlap the horizontal segment of the i-th net on the j-th track of the k-th layer.

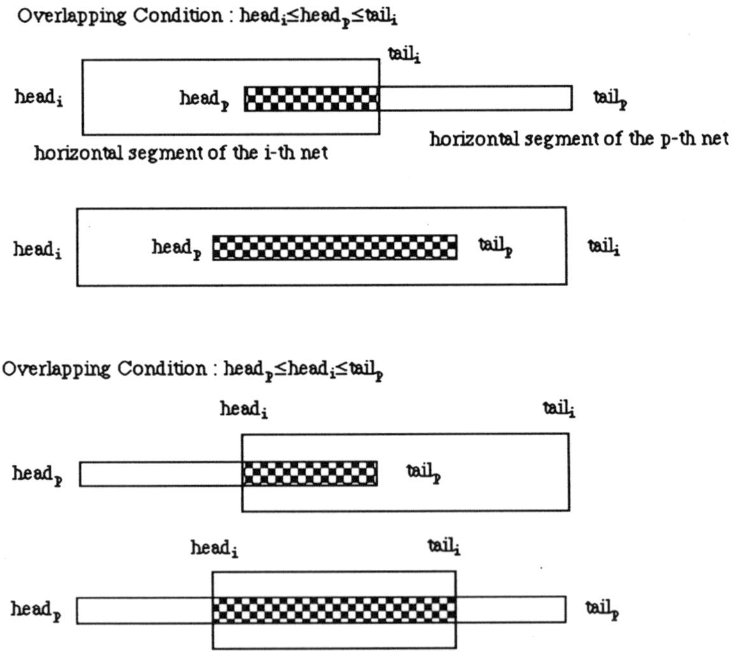

Fig. 6 Overlapping conditions for horizontal segments.

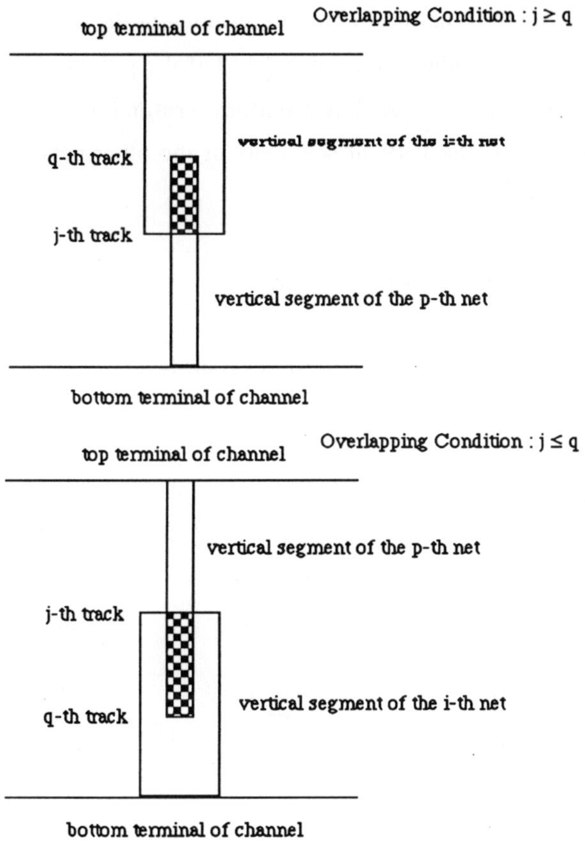

Fig. 7 Overlapping conditions for vertical segments.

Fig. 7 shows the overlapping conditions for the vertical segments of the nets where the i-th net and the p-th net have the terminals on the opposite sides of the same column. The vertical overlapping conditions for the i-th-net-j-th-track-k-th-layer neuron are given by:

$$\sum_{\substack{p=0\\p\neq i}}^{n-1} T_{ip}\sum_{q=0}^{j} V_{pqk}+\sum_{\substack{p=0\\p\neq i}}^{n-1} B_{ip}\sum_{q=j}^{m-1} V_{pqk}$$

where T_{ip} is 1 if the p-th net has a bottom terminal at the column at which the i-th net has a top terminal, 0 otherwise, and B_{ip} is 1 if the p-th net has a top terminal at the column at which the i-th net has a bottom terminal, 0 otherwise. The vertical condition is nonzero if the vertical segments of the other nets overlap the vertical segments of the i-th net.

THE MOTION EQUATION OF U_{ijk}

The motion equation of the i-th-net-j-th-track-k-th-layer neuron for a n-net-m-track-2k-layer problem is given by:

$$\frac{dU_{ijk}}{dt}=-A\left(\sum_{q=0}^{m-1}\sum_{r=0}^{l-1} V_{iqr}-1\right)-B\left(\sum_{\substack{p=0\\ p\neq i\\ head_i\leq head_p\leq tail_i}}^{n-1} V_{pjk}+\sum_{\substack{p=0\\ p\neq i\\ head_p\leq head_i\leq tail_p}}^{n-1} V_{pjk}\right)$$

$$-B\left(\sum_{\substack{p=0\\ p\neq i}}^{n-1} T_{ip}\sum_{q=0}^{j} V_{pqk}+\sum_{\substack{p=0\\ p\neq i}}^{n-1} B_{ip}\sum_{q=j}^{m-1} V_{pqk}\right)+Ch\left(\sum_{q=0}^{m-1}\sum_{r=0}^{l-1} V_{iqr}\right)$$

The first term (A-term) forces only one output among the $l\times m$ neurons to be nonzero where the i-th net is assigned. The second and third terms (B-terms) perform the inhibitory forces. The B-terms discourage the output of the ijk-th neuron to be nonzero if the other nets overlap with the i-th net. The last term (C-term) provides the hill-climbing which allows the state of the system to escape from the local minimum and to converge on the global minimum. The C-term encourages the output of the ijk-th neuron to be nonzero if the output of all the neurons for the i-th net is zero. The function h(x) is 1 if x=0, 0 otherwise. A, B, and C are constant coefficients, and t is for time.

THE MAIN FLOW OF THE ALGORITHM

Using the first order Euler method, the main flow of the algorithm is described as following.

0. Set t = 0, A = B = 1, C = 10, $U_{max} = 20$, $U_{min} = -20$, and $T_{max} = 500$.

1. All $U_{ijk}(t)$ (for i = 0,..., n-1, j = 0,..., m-1, and k = 0, .., l-1) are uniformly randomized between 0 and U_{min}, and assign 0 to all $V_{ijk}(t)$ (for i = 0,..., n-1, j = 0,..., m-1, and k = 0, .., l-1).

2. Use the motion equation to compute all $\frac{dU_{ijk}(t)}{dt}$.

3. Compute all $U_{ijk}(t+1)$.

$$U_{ijk}(t+1)=U_{ijk}(t)+\frac{dU_{ijk}(t)}{dt}$$

4. Check all $U_{ijk}(t+1)$.

$$U_{ijk}(t+1) = U_{min}, \text{ if } U_{ijk}(t+1) < U_{min}.$$
$$= U_{max}, \text{ if } U_{ijk}(t+1) > U_{max}.$$

5. Compute all $V_{ijk}(t+1)$.

6. If all nets are embedded without conflicts or t = T_{max} then terminate this procedure else increment t by 1 and return to step 2.

5.4 TARGET MACHINE

The algorithm was implemented on a multiprocessor testbed ATTEMPT-0 (A Typical Testing Environment of MultiProcessor sysTem version 0) (Amano H., Terasawa T. and Kudoh T. 1989) developed at Keio University. In this system, ten processors are connected with a shared memory through the IEEE Futurebus. Each processor board consists of a CPU/FPU (68030/68882), a 4MB local memory system and a write through snooping cache (Fig. 8), and a special communication mechanism called the synchronizer which is provided for synchronization and high speed data multicasting. Instead of using a dedicated shared memory board, the local memory of a processor board can be used as a shared memory. The multiprocessor system used in this experiment comprises eight processors each of which provides the 4MB local memory and the 8MB shared memory because two processor boards are used only as the shared memory.

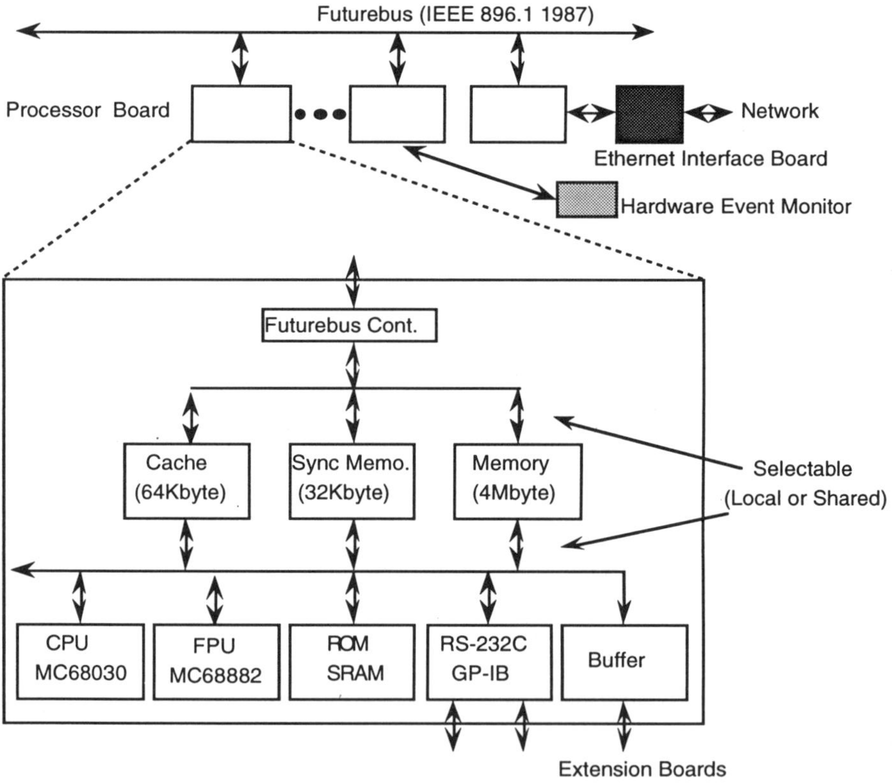

Fig. 8 A target multiprocessor : ATTEMPT-0.

5.5 IMPLEMENTATION

Although a lot of parallel channel routing algorithms have been proposed and implemented on multiprocessor systems (Chang S. C. and J\. aJ\. a J\. 1988a) (Chang S. C. and J\. aJ\. a J. 1988b) (Chang S. C. and J\. aJ\. a J. 1988c) (Krishnamurthy S. and J\. aJ\. a J. 1988) (Iosupovici A. 1986) (Rose J. 1990) (Brouwer R. J. and Banerjee P. 1988) (Watanabe T., Kitazawa H., and Sugiyama Y. 1987) (Brower R. J. and Banerjee P. 1990), any channel routing algorithm which can treat more than four-layer problems and/or which is based on a neural network model has not been implemented on a multiprocessor system. In the algorithm, each neuron can be a unit of parallel computing. However, when neurons for a net are assigned on different processors, a large amount of communication is required between them. In order to alleviate it, a net is chosen as a unit of parallel computing and all neurons for a net are assigned on the same processor. In an iteration step of the Euler method, every processor updates variables U and V until all processors recognize the convergence of the system. Our target multiprocessor system, ATTEMPT-0 provides both local and shared memories. To make the best use of the structure of ATTEMPT-0, variables which are not needed to be shared must be located on the local memory as much as possible. All the variables V are needed for updating each valuable U while only the variables U of a net are needed for updating the variables V of the net. Thus, the variables V should be located on the shared memory while the variables U are located on the local memory To maintain the load balancing between processors, we used the following two methods:

Simple method:

Nets are assigned into processors so as to minimize the difference between the number of nets per processor.

Modified method:

Nets are assigned into processors so as to minimize the difference not only between the number of nets per processor but also on the total number of terminals in nets per processor. Although the load unbalance is reduced, it takes a setup time to assign nets into processors with this method.

In order to keep sequential consistency, processors must be synchronized after step 1, after step 4, after step 5, and at step 6 in the main flow of the algorithm in section 3-E.

5.6 EXPERIMENTAL RESULTS

The benchmark problems in (Yoshimura T. and Kuh E. S. 1982) are often used for evaluation of channel routing algorithms. Four problems in those benchmark problems, example 3a, example 3c, example 5, and Deutsch's difficult example are used in this experiment. Table 1 shows the number of nets and the total number of terminals in all nets for each problems.

Table 1 Benchmark problems.

problem	No. of terminals	No. of nets
example 3a	45	120
example 3c	54	152
example 5	61	168
Deutsch's	72	302

In order to evaluate the execution time under the same condition, the number of layers is set to be four and the number of iteration steps for updating variables U and V is set to be a constant 100. This constant number is the average number of iteration steps until the neural network model converges on a solution.

EXECUTION SPEED

Fig. 9 and Fig. 10 show the improvement of execution speed by using the simple method and the modified method respectively. The execution speed is compared with that of the sequential version executed with one processor. Between 5.0 and 6.0 times improvement is achieved with 8 processors by using the simple method. By using the modified method, it is improved to be between 6.0 and 6.5 times with 8 processors, and the difference of the execution speed between problems is reduced. Even in Deutsch's difficult example, the speed up is linear.

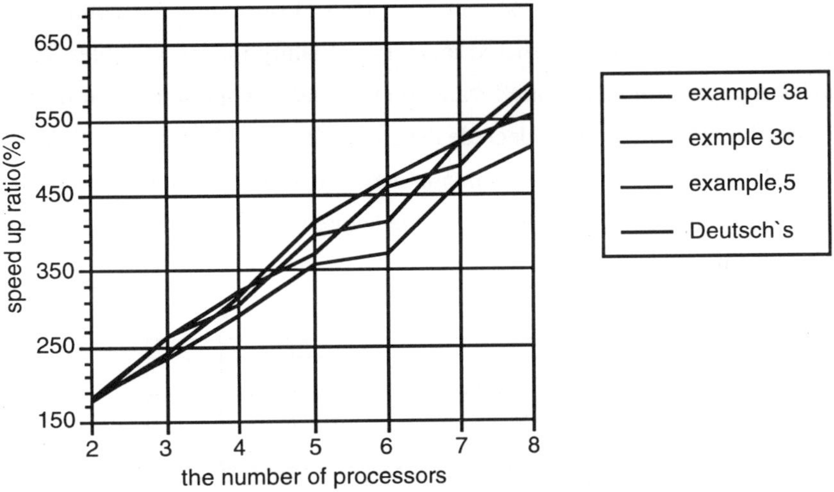

Fig.9 The number of processors v.s. the speed up rate (Simple Method).

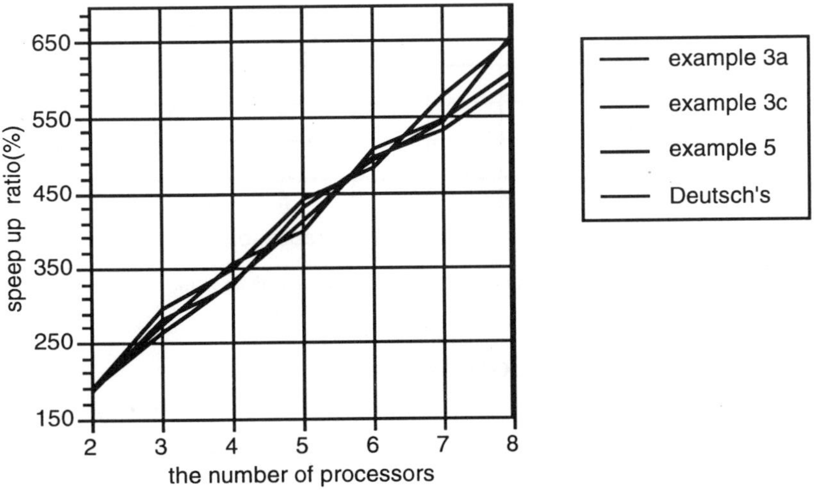

Fig. 10 The number of processors v.s. the speed up rate(Modified Method).

CACHE HIT RATIO

Fig. 11 shows the number of processors v.s. the cache hit ratio in Deutsch's difficult example. The cache hit ratio is kept high (Fig.11) because the communication between processors using the shared memory is mainly for the variables V. Shared variables for synchronization are located on the special communication mechanism called synchronizer, and do not degrade the cache hit ratio. Though the ratio decreases with the number of processors, maintains larger than 97\% even in the worst case because of the continuous data access by the nature of the algorithm. The high cache hit ratio contributes high speed execution.

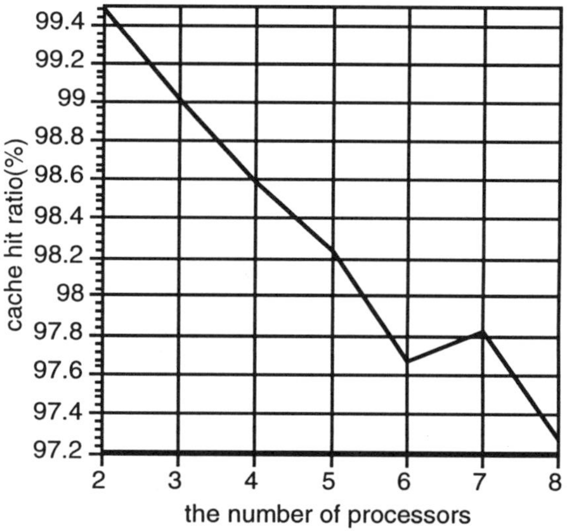

Fig. 11 The number of processors v.s. cache hit ratio.

COMMUNICATION OVERHEAD

Fig. 12 shows the ratio of the communication overhead to the total execution time in Deutsch's difficult example. The overhead of the synchronization (sync) does not appear in the figure because processors are synchronized only a few times in each iteration. The overhead of writing data (write) is small while the overhead of the cache block replacing (replace) dominates the communication overhead. It is caused by the large cache block size (64byte) and relatively slow block data transfer on the bus of ATTEMPT-0. Since the local memory is effectively utilized in the implementation, the ratio of communication overhead is kept low (1.1\% in maximum).

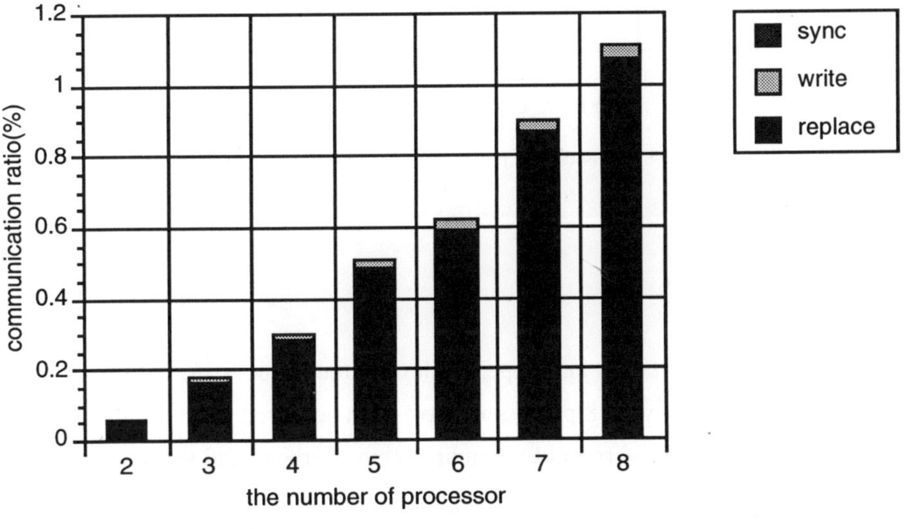

Fig. 12 The number of processors v.s. communication overhead.

Fig. 13 shows the ratio of waiting time caused by the bus congestion to the total execution time in Deutsch's difficult example. Although it increases with the number of processors, it is kept less than 1.1\%.

More than 8 processors can be used effectively for better performance because cache hit ratio is kept high, and both the ratio of communication overhead and the ratio of bus waiting time are kept low.

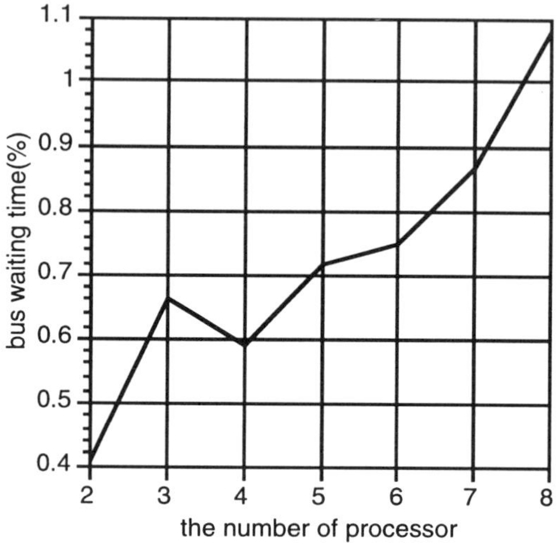

Fig. 13 The number of processors v.s. bus congestion ratio.

5.7 CONCLUSION

Through the implementation and experiment on a multiprocessor, between 6.0 and 6.5 times improvement with 8 processors have been achieved. Because the cache hit ratio is kept high while the ratio of communication overhead and waiting time to the total execution time is kept low. The algorithm was implemented on a relatively small multiprocessor system. However, considering the experimental results, implementation in this paper can be effectively extended for a larger scale multiprocessor system even without the shared memory.

5.8 ACKNOWLEDGEMENT

We would like to thank Junichi Yamato for his assistance in analysis about the experimental results and Nobuo Funabiki for his advice in converting the algorithm from an sequential machine into a multiprocessor system.

5.9 REFERENCES

Amano H., Terasawa T. and Kudoh T., (1989), Cache with synchronization mechanism, Proc. IFIP Congress89, pp.1001-1006.

Brouwer R. J. and Banerjee P., (1988), A parallel simulated annealing algorithm for channel routing on a hypercube multiprocessor, IEEE Int. Conf. Comput. Design, pp.4-7.

Brouwer R. J. and Banerjee P., (1990), PHIGURE: a parallel hierarchical global router, Proc. 27th Design Automation Conf., pp.650-653.

Burstein M., and Pelavin R., (1983), Hierarchical wire routing, IEEE Trans. CAD, Vol.CAD-2, No.4, pp.223-234.

Chang S.C. and J\. aJ\. a J., (1988a), Parallel algorithms for channel routing in the knock-knee model, Proc. Int. Conf. Parallel Processing, pp.18-25.

Chang S.C. and J\. aJ\. a J., (1988b), Parallel algorithms for river routing, Proc. Int. Conf. Parallel Processing}, pp.9-13.

Chang S.C. and J\. aJ\. a J., (1988c), Optimal mesh algorithms for VLSI routing, Proc. 2nd Symp. Frontiers Massively Parallel Comput., pp.18-25.

Chen Y.K. and M.L.Liu, (1984), Three-layer channel routing, IEEE Trans. CAD, Vol.CAD-3, No.2, pp.156-163.

Cong J. and Liu C. L., (1990), Over-the-cell channel routing, IEEE Trans. CAD, Vol.CAD-9, No.4, pp.408-418.

Cong J. and Liu C. L., "Over-the-cell channel routing," Proc. ICCAD-88, pp.80-83, 1988.

Cong J., Wong D. F., and Liu C. L., (1987), A new approach to the three layer channel routing problem, Proc. ICCAD-87, pp.378-381.

Cong J., Wong D. F., and Liu C. L., (1988), A new approach to three- or four-layer channel routing, IEEE Trans. CAD, Vol.CAD-7, No.10, pp.1094-1104.

Deutsch D. N. and P. Glick, (1980), An over-the-cell router, Proc. 17th IEEE/ACM

Krohn H. E., (1983), An over-cell gate array channel router, Proc. 20th Design Automation Conf., pp.665-670.

Deutsch D. N., (1976), A dogleg channel router, Proc. 13th Design Automation Conf., pp.425-433.

Deutsch D. N., (1989), Two new and more difficult channel routing problems, IEEE Trans. CAD, Vol.CAD-8, No.4, pp.448.

Dolev D. et al., (1981), Optimal wiring between rectangles, Proc. 13th Annual ACM Symp. Theory of Comput., pp.312-317.

Funabiki N. and Takefuji Y., (1992), A parallel algorithm for channel routing problems, IEEE Trans. CAD, Vol.CAD-11, No.4, pp.464-474.

Gudmundsson G. and Ntafos S., (1987), Channel routing with superterminals, Proc. 25th Allerton Conf. on Comput., Control and Communication, pp.375-376.

Hashimoto A. and Stevens J., (1971), Wire routing by optimizing channel assignment with large apertures, Proc. 8th Design Automation Workshop, pp.155-

Kernighan B. W., Schweikert D. G., (1973), An optimum channel-routing algorithm for polycell layouts of integrated circuits, Proc. 10th Design Automation Workshop, pp.50-59.

Iosupovici A., (1986), A class of array architectures for hardware grid routers, IEEE Trans. CAD, Vol.CAD-5, No.2, pp.245-255.

Joobbani R. and Siewiorek D. P., (1985), Weaver: A knowledge-based routing expert, Proc. 22nd Design Automation Conf., pp.266-272.

Krishnamurthy S. and J\. aJ\. a J., (1988), Provably good parallel algorithms for channel routing of multi-terminal nets, Proc. 2nd Symp. Frontiers Massively Parallel Comput., pp.177-180.

LaPaugh A. S., (1980), Algorithms for integrated circuits layout: An analytic approach, Ph.D. dissertation, M.I.T. Lab. Computer Science.

Leong H. W. and Liu C. L., (1983), A new channel routing problem, Proc. 20th Design Automation Conf., pp.584-590.

Leong H. W., Wong D. F., and Liu C. L., (1985), A simulated-annealing channel router," Proc. ICCAD-85, pp.226-228.

Reed J., Sangiovanni-Vincentelli A., and Santomauro M., (1985), A new symbolic channel router: YACR2, IEEE Trans. CAD, Vol.CAD-4, No.3, pp.208-219.

Rivest R. L. and Fiduccia C. M., (1982), A greedy channel router, Proc. 19th Design Automation Conf., pp.418-424.

Rose J., (1990), Parallel global routing for standard cells, IEEE Trans. CAD, Vol.CAD-9, No.10, pp.1085-1095.

Sahni S. and Bhatt A., (1980), The complexity of design automation problems, Proc. 17th Design Automation Conf., pp.402-411.

Shiraishi Y. and Sakemi J., (1987), A permeation router, IEEE Trans. CAD, Vol.CAD-6, No.3, pp.462-471.

Szymanski T. G., (1985), Dogleg channel routing is NP-complete, IEEE Trans. CAD, Vol.CAD-4, No.1, pp.31-41.

Takefuji Y., (1992), Neural network parallel computing, Kluwer Academic Publishers.

Watanabe T., Kitazawa H., and Sugiyama Y., (1987), A parallel adaptable routing algorithm and its implementation on a two-dimensional array processor, IEEE Trans. CAD, Vol.CAD-6, No.2.

Yoshimura T. and Kuh E. S., (1982), Efficient algorithms for channel routing, IEEE Trans. CAD, Vol.CAD-1, No.1, pp.25-35.

Chapter 6

JOB-SHOP SCHEDULING

SIMON Y. FOO
Department of Electrical Engineering
FAMU/FSU College of Engineering
Florida State University
Tallahassee, Florida 32316, USA

YOSHIYASU TAKEFUJI
Dept. of Electrical Engineering and Applied Physics
Case Western Reserve University
Cleveland, OH 44106, USA
and
Faculty of Environmental Information
Keio University, Fujisawa 252, Japan

The Tank-Hopfield linear programming network is modified to solve job-shop scheduling, a classical optimization problem. Using a linear energy function, our approach avoids the traditional problems associated with most Hopfield networks using quadratic energy functions. Although our approach requires more hardware (in terms of processing elements and resistive interconnects) than a recent approach by Zhou et. al. (1991), the neurons in the modified Tank-Hopfield network do not perform extensive calculations unlike those described by Zhou et. al.

6.1 INTRODUCTION

Hopfield neural networks have been successfully applied to solving a variety of interesting problems such as the traveling-salesman problem, analog-to-digital conversion [1], resource allocation, tiling problem, k-colorability problem, and many others as described by Takefuji [5]. Recent criticisms of the Hopfield networks such as nonconvergence of the network to valid solutions, inability to locate the global minimum, and poor scaling properties, are due to the use of quadratic energy functions as pointed out by Zhou et. al. [2]. This work as an extension of our earlier results reported in [4] uses a modified Tank- Hopfield linear programming network with a linear energy function which overcomes the shortcomings of the traditional quadratic energy functions.

6.2 JOB-SHOP SCHEDULING

Job-shop scheduling belongs to the large class of NP-complete (nondeterministic polynomial time complete) problems. An NP-complete problem exhibits an exponential growth in the computation time as the size of the problem increases linearly, and is commonly referred to as a "hard" problem. In general, an NP-complete problem is one which, for some input of size N, takes a time proportional to at least 2^N. On the contrary, a non-NP (deterministic polynomial time) problem shows a linear growth in the computation time as the size of the problem increases linearly. An example of non-NP problems is the application of an insertion sort routine on a table of N entries, where the computation time is proportional to N^2. NP-complete problems occur naturally in many situations. For example, there are more than $N!$ ways to schedule N courses to various classrooms and instructors, which means that determining a complete schedule of a large number of classes pleasing to both students and professors can become a formidable task!

In general, the job-shop scheduling problem can be stated as follows: Given n

jobs that have to be processed on *m* machines in a prescribed order under certain restrictive assumptions, what is then the optimal order in which each machine handles the jobs? In short, job-shop scheduling is a resource allocation problem. The resources are called *machines* and basic tasks are called *jobs*. Each job may consist of several subtasks referred to as *operations* that are interrelated by precedence restrictions. A close resemblance to the job-shop problem is processor scheduling in a multiprocessor computer system where the problem definition is: Given a deadline and a set of tasks of varying length to be performed on two identical processors, can the tasks be arranged so that the deadline is met?

In a general job-shop problem, each operation is described by a triplet (i,j,k), i.e., operation j of job i is to be executed on machine k. Assuming there are m machines, then each job has exactly m operations with exactly one operation on each machine. If there are n jobs, then each machine must perform n operations, which means that the number of possible sequences is therefore n! for each machine. If the sequences on each machine are independent, there are $(n!)^m$ schedules. However, since each job consists of several operations with a linear precedence structure, many of these schedules will have conflicts and therefore are invalid solutions.

An example 2-job 3-machine job-shop problem is shown by a Gantt chart in Figure 1(a). Each job-operation-machine triplet (i,j,k) is represented by a block; the length of block is equal to the processing time t_{ijk} required to perform the operation while the numbers on the horizontal axis represent completion times. Note that the processing order of each job by all machines and processing time of each operation are assumed known and fixed (static).

A feasible schedule is one where all operations of each job can be placed on one time axis in precedence order and without overlap. In principle, there are infinitely many feasible schedules for a given job-shop problem since an arbitrary amount of idle time can be inserted at any machine between pairs of operations. Superfluous idle time exists in a schedule if an operation can be processed earlier in time without altering the operation sequences on any machine. In a nondelay schedule, no machine is kept idle at a time when it could begin processing some operation. Although nondelay schedules can be expected to provide very good solutions, there is no guarantee that they provide the optimum solutions. A job-shop problem is completely solved if the starting times of all operations are determined, and the precedence relationships between the operations are not violated. In general, the optimality criteria for machine scheduling can be classified into various groups. These measures of performance include criteria based on completion-dates (i.e., the time at which the last job-operation is completed), flow-times (i.e., the amount of time the job spends in the shop), and makespan (i.e., the total elapsed time required to process all of a given set of jobs). In many cases, previous researchers have used minimization of makespan as the objective function. The optimum solution to the 2-job 3-machine problem is shown in Figure 1(b).

6.3 PROBLEM FORMULATIONS

Several integer linear programming formulations of the machine sequencing problem have been proposed in the past but no attempt has been made to identify and exploit any special structure of this integer linear programming technique. Recently, we [4] proposed a mixed integer linear approach to solving job-shop scheduling. In our approach, the determination of an optimal job-shop schedule can be formulated as a linear programming with integer adjustments to minimize the starting times of all jobs subject to a set of precedence constraints and restrictive assumptions.

Let S_{ik} denote the starting time of job i on machine k, and t_{ijk} the processing time for operation (i,j,k). Assuming operation $(i,j\text{-}1,h)$ precedes (i,j,k), then the inequalities representing precedence constraints in order for a set of S_{ik} to be feasible are:

$$S_{ik} - S_{ih} \geq t_{i,j-1,h} \qquad 1 \leq j \leq m, \qquad 1 \leq i \leq n \tag{1}$$

where m and n are the number of machines and number of jobs, respectively. The condition that all starting times must be positive results in the constraint:

$$S_{ik} \geq 0 \qquad 1 \leq i \leq n \tag{2}$$

It is also necessary to ensure that no two operations are processed simultaneously by the same machine at the same time. For example, if job i precedes job p on machine k (i.e., operation (p,q,k) starts after the completion of (i,j,k)), then

$$S_{pk} - S_{ik} \geq t_{ijk}$$

On the other hand, if job p precedes job i on machine k, it is also necessary that

$$S_{ik} - S_{pk} \geq t_{pqk}$$

Using a "zero-one" variable y_{ipk} to specify the operation sequence, i.e., $y_{ipk} = 1$ if job i precedes job p on machine k, and $y_{ipk} = 0$ otherwise, the above constraints become:

$$S_{pk} - S_{ik} + H*(1 - y_{ipk}) \geq t_{ijk} \tag{3}$$

$$S_{ik} - S_{pk} + H*y_{ipk} \geq t_{pqk} \tag{4}$$

where constant H represents an arbitrary positive number greater than the maximum value of all processing times t_{ijk}'s such that the constraints (3) and (4) are satisfied.

Therefore, the entire job-shop problem formulation can be summarized as minimizing the cost function

$$\sum_{i=1}^{n} S_{ik}$$

subject to

$$
\begin{array}{ll}
S_{ik} - S_{ih} \geq t_{i,j-1,h} & 1 \leq j \leq m, \quad 1 \leq i \leq n \\
S_{pk} - S_{ik} + H*(1 - y_{ipk}) \geq t_{ijk} & i \geq 1, \quad p \leq n, 1 \leq k \leq m \\
S_{ik} - S_{pk} + H*y_{ipk} \geq t_{pqk} & i \geq 1, \quad p \leq n, 1 \leq k \leq m \\
S_{ik} \geq 0 & 1 \leq i \leq n
\end{array}
$$

where k_i is the machine which the last operation of job i is assigned. There are mn constraints of type (1) or type (2), and $mn(n-1)$ constraints of type (3) or type (4), giving a total number of mn^2 constraints. There are also mn number of S_{ik}'s and $mn(n-1)/2$ number of y_{ipk}'s, resulting a total number of $mn(n+1)/2$ variables. For a 2-job 3-machine problem, there are a total of 12 inequalities with 9 variables in the formulation, i.e.,

$$
\begin{array}{l}
S_{11} \geq 0 \\
S_{12} - S_{11} \geq t_{111} \\
S_{13} - S_{12} \geq t_{122} \\
S_{21} - S_{23} \geq t_{213} \\
S_{22} - S_{21} \geq t_{221} \\
S_{23} \geq 0 \\
S_{21} - S_{11} + H*(1 - y_{121}) \geq t_{111} \\
S_{11} - S_{21} + H*y_{121} \geq t_{221} \\
S_{22} - S_{12} + H*(1 - y_{122}) \geq t_{122} \\
S_{12} - S_{22} + H*y_{122} \geq t_{232} \\
S_{23} - S_{13} + H*(1 - y_{123}) \geq t_{133} \\
S_{13} - S_{23} + H*y_{123} \geq t_{213}
\end{array}
$$

6.4 NETWORK ARCHITECTURE

The analog Tank and Hopfield linear programming network [1] seeks to minimize a cost function

$$F(\mathbf{A},\mathbf{V}) = \mathbf{A}\mathbf{V}$$

where **A** is row vector array of N coefficients for N variables which are components of column vector **V**. This linear cost function is subject to a set of M linear constraints among the N variables:

$$\mathbf{D}_j \bullet \mathbf{V} \geq \mathbf{B}_j \qquad j = 1,..,M$$

where matrix $\mathbf{D}_j$ contain the N variable coefficients of constraint equation j and column vector $\mathbf{B}_j$ are the bounds.

Our proposed network for solving mixed integer-linear programming problems is based on a slightly modified Tank and Hopfield network, as shown in Figure 2. The difference is the addition of nonlinear h-amplifiers. The output voltage V_i of linear g-amplifier represents the starting time of each operation after minimization, while output Y_i of nonlinear h-amplifier represents the zero-one variable. The components of **A** are proportional to the input currents fed into the g- and h-amplifiers. Each g- and h-amplifier has an input time constant $\rho_i C_i$ which acts as a memory element. The g-amplifier circuit can be implemented as a noninverting amplifier which obeys a linear relationship, as shown in Figure 3. The gain of the g-amplifier is adjusted by varying R_s or R_f. The h-amplifier is simply a high-gain analog comparator as shown in Figure 4. The neutral position of the step function is adjusted by V_{ref}. Description of other analog components such as variable resistors, sigmoid amplifier circuits for implementing neural networks can be found in [3].

The M outputs (ψ_j) of the f-amplifers represent constraint satisfaction. This set of amplifiers have a nonlinear transfer function

$$\psi_j = f(U_j) = \begin{cases} 0 & U_j \geq 0 \\ -U_j & U_j < 0 \end{cases}$$

where

$$\mathbf{U}_j = \mathbf{D}_j \bullet \mathbf{V} - \mathbf{B}_j$$

i.e., if $U_j < 0$ the output ψ_j of the f-amplifier will be a large positive value indicating a violation of the j-th constraint. When this happens, a current of value $(\psi_j D_{ji})$ will be fed back into the inputs of the g- and h-amplifiers, causing the outputs V_i to oscillate between +Vcc and -Vcc, where Vcc is the power supply voltage of the operational amplifiers. If $U_j \geq 0$ the output ψ_j of the f-amplifier is zero corresponding to a valid solution and no current will be fed back to the g- and h-amplifiers. As a result, the output voltages of the g- and h-amplifers will be stable and directly proportional to the input bias currents A_i's.

The nonlinear f-amplifier can be implemented using a precision inverting half-wave rectifier as shown in Figure 5. The rectifier circuit operates in the following manner: For positive V_I, diode D_2 conducts and closes the negative feedback loop around the op-amp. A virtual ground appears at the inverting input terminal and the op-amp output is clamped at –0.7V. This negative voltage keeps D_1 in cut-off which means no current flows through feedback resistance R_2. Hence the rectifier output voltage is zero. When V_I goes negative, the voltage at the op-amp output terminal becomes positive, causing D_2 to be reversed-biased. Since the inverting input terminal of the op-amp always appears as virtual ground, therefore diode D_1 is forward-biased and thus conducts through R_2 and establishing a negative feedback loop around the op-amp. The current through the feedback resistance R_2 is equal to the current through input resistance R_1. Therefore, if R_1=R_2, then the output voltage of the rectifier is:

$V_O = -V_I \quad \text{for } V_I \leq 0$

The complete transfer function of the precision half-wave rectifier is shown in Figure #(b). Notice that the slope of the characteristic can be set to any desired value by selecting the appropriate values for R_1 and R_2.

Since the currents generated by the g- and h-amplifiers are summed at the inputs of the f-amplifiers, it is necessary that the time constants ($\rho_i C_i$) at the inputs of g- and h-amplifiers be matched. If we assume the response time of the f-amplifier is negligible compared to the g- and h-amplifiers, then the dynamics of the g-amplifier can be described by:

$$C_i \frac{dU_i}{dt} = -A_i - \frac{U_i}{R_i} - \sum_j^{M} D_{ji}\, f(D_j \bullet V - B_j) \tag{5}$$

where $M = mn^2$ (i.e., total number of constraints). Similarly, the differential equation describing the behavior of the h-amplifier is:

$$C_i \frac{dX_i}{dt} = -A_i - \frac{X_i}{R_i} - \sum_j^{M} D_{ji}\, f(D_j \bullet Y - B_j) \tag{6}$$

Consider an energy function for the entire circuit of the form:

$$E = \sum_i^{P} A_i V_i + \sum_i^{P} \frac{1}{R_i} \int_0^{V_i} g^{-1}(V)\,dV + \sum_i^{P}\sum_j^{M} \int f(z)\,dz + \sum_i^{Q} \frac{1}{R_i} \int_0^{Y_i} h^{-1}(Y)\,dY + \sum_i^{Q}\sum_j^{M} \int f(w)\,dw$$

where

$$U = g^{-1}(V)$$
$$X = h^{-1}(Y)$$
$$z = D_j \bullet V - B_j$$
$$w = D_j \bullet Y - B_j$$

$P = mn$ (i.e., total number of S_{ik}'s)

$Q = mn(n-1)/2$ (i.e., total number of y_{ipk}'s)

It can be easily shown that E is always a decreasing energy function which seeks out a minima in E and stops. The proof is as follows:

$$\begin{aligned}\frac{dE}{dt} &= \sum_i^P A_i \frac{dV_i}{dt} + \sum_i^P \frac{1}{R_i} U_i \frac{dV_i}{dt} + \sum_i^P \sum_j^M f(z) \frac{dz}{dt} + \sum_i^Q \frac{1}{R_i} X_i \frac{dY_i}{dt} + \sum_i^Q \sum_j^M f(w) \frac{dw}{dt} \\ &= \sum_i^P A_i \frac{dV_i}{dt} + \sum_i^P \frac{U_i}{R_i} \frac{dV_i}{dt} + \sum_i^P \sum_j^M f(D_j \bullet V - B_j) D_{ji} \frac{dV_i}{dt} + \sum_i^Q \frac{X_i}{R_i} \frac{dY_i}{dt} + \sum_i^Q \sum_j^M f(D_j \bullet Y - B_j) D_{ji} \frac{dY_i}{dt} \\ &= \sum_i^P \frac{dV_i}{dt} \left\{ -C_i \frac{dU_i}{dt} \right\} + \sum_i^Q \frac{dY_i}{dt} \left\{ -C_i \frac{dX_i}{dt} \right\} \\ &= -\sum_i^P C_i g^{-1\prime}(V_i) \left(\frac{dV_i}{dt} \right)^2 - \sum_i^Q C_i h^{-1\prime}(Y_i) \left(\frac{dY_i}{dt} \right)^2\end{aligned}$$

Since C_i is positive, and $g^{-1\prime}(V_i)$ and $h^{-1\prime}(Y_i)$ are monotone increasing and step functions, respectively, it follows that $dE/dt \leq 0$ with $dE/dt = 0$ if $dV_i/dt = dY_i/dt = 0$ for all i. In other words, the time evolution of the network is a motion in state space which seeks out a minima in E and stops.

6.5 NETWORK SIMULATIONS

To illustrate how the modified Tank and Hopfield network works, consider the formulation of an example 2-job 3-machine problem shown in Figure 1(a). Let x_1 through x_6 uniquely identify the starting times of the operations while x_7 through x_9 represent the zero-one variables, and V_i represents the output voltage of the g- and h-amplifiers corresponding to variable x_i. Choosing constant H=10 (i.e., any number greater than the maximum processing time value of 9), the complete set of inequalities are:

$$x_1 \geq 0$$
$$-x_1 + x_2 \geq 5$$
$$-x_2 + x_3 \geq 8$$
$$x_4 - x_6 \geq 7$$
$$-x_4 + x_5 \geq 3$$
$$x_6 \geq 0$$
$$-x_1 + x_4 - 10x_7 \geq -5$$
$$x_1 - x_4 + 10x_7 \geq 3$$
$$-x_2 + x_5 - 10x_8 \geq -2$$
$$x_2 - x_5 + 10x_8 \geq 9$$
$$-x_3 + x_6 - 10x_9 \geq -8$$
$$x_3 - x_6 + 10x_9 \geq 7$$

Then the corresponding $\mathbf{D_j}$ matrix is:

$$\begin{bmatrix}
1 & 0 & 0 & 0 & 0 & 0 & 0 & 0 & 0 \\
-1 & 1 & 0 & 0 & 0 & 0 & 0 & 0 & 0 \\
0 & -1 & 1 & 0 & 0 & 0 & 0 & 0 & 0 \\
0 & 0 & 0 & 1 & 0 & -1 & 0 & 0 & 0 \\
0 & 0 & 0 & -1 & 1 & 0 & 0 & 0 & 0 \\
0 & 0 & 0 & 0 & 0 & 1 & 0 & 0 & 0 \\
-1 & 0 & 0 & 1 & 0 & 0 & -10 & 0 & 0 \\
1 & 0 & 0 & -1 & 0 & 0 & 10 & 0 & 0 \\
0 & -1 & 0 & 0 & 1 & 0 & 0 & -10 & 0 \\
0 & 1 & 0 & 0 & -1 & 0 & 0 & 10 & 0 \\
0 & 0 & -1 & 0 & 0 & 1 & 0 & 0 & -10 \\
0 & 0 & 1 & 0 & 0 & -1 & 0 & 0 & 10
\end{bmatrix}$$

Using 1kΩ as base unit, the resistor network for implementing $\mathbf{D_j}$ becomes:

$$\begin{bmatrix}
1K & \infty & \infty & \infty & \infty & \infty & \infty & \infty & \infty \\
1K & 1K & \infty & \infty & \infty & \infty & \infty & \infty & \infty \\
\infty & 1K & 1K & \infty & \infty & \infty & \infty & \infty & \infty \\
\infty & \infty & \infty & 1K & \infty & 1K & \infty & \infty & \infty \\
\infty & \infty & \infty & 1K & 1K & \infty & \infty & \infty & \infty \\
\infty & \infty & \infty & \infty & \infty & 1K & \infty & \infty & \infty \\
1K & \infty & \infty & 1K & \infty & \infty & 0.1K & \infty & \infty \\
1K & \infty & \infty & 1K & \infty & \infty & 0.1K & \infty & \infty \\
\infty & 1K & \infty & \infty & 1K & \infty & \infty & 0.1K & \infty \\
\infty & 1K & \infty & \infty & 1K & \infty & \infty & 0.1K & \infty \\
\infty & \infty & 1K & \infty & \infty & 1K & \infty & \infty & 0.1K \\
\infty & \infty & 1K & \infty & \infty & 1K & \infty & \infty & 0.1K
\end{bmatrix}$$

where "∞" denotes "no connection". If equal weight is placed on all production costs such as operation costs, machine idle costs, etc., then the input bias currents $-A_i$'s can be arbitrarily set to −1 mA. If an external potential of −1 volt is applied to provide the $-A_i$'s, then the set of input resistances is [1kΩ, 1kΩ, .., 1kΩ].

Let each unit of processing time be represented by 1 mA of current. Then the current vector representing the bounds of constraint equations is $-\mathbf{B_j}$ = [0, −5 mA, −8 mA, −7 mA, −3 mA, 0, 5 mA, −3 mA, 2 mA, −9 mA, 8 mA, −7 mA]. Assuming the set of resistor values for implementing $-\mathbf{B_j}$ is [100Ω, 100Ω, .., 100Ω], then the input potentials to the set of resistors must be [0, −500mV, −800 mV, −700 mV, −300 mV, 0, +500 mV, −300 mV, +200 mV, −900 mV, +800 mV, −700 mV]. The complete circuit for solving the 2-job 3-machine problem is shown in Figure 6.

When the network reaches stable states, the output voltages of g- and h-amplifiers are measured. The results are (in mV):

$V_1 = 0.0$ $V_2 = 0.5$ $V_3 = 1.3$ $V_4 = 0.7$ $V_5 = 1.3$ $V_6 = 0.0$

$Y_1 = 1.0$ $Y_2 = 1.0$ $Y_3 = 0.0$

where V_i's are outputs of the linear g-amplifiers representing the starting times, and Y_i's are outputs of the h-amplifiers representing the zero-one variables. By normalization, the starting times for the operations are:

$S_{11} = 0$ $S_{12} = 5$ $S_{13} = 13$ $S_{21} = 7$ $S_{22} = 13$ $S_{23} = 0$

Figure 1(b) shows the Gantt chart of the job-shop schedule constructed from the above results which turns out to be an optimum schedule. Another simulation run was performed based on a 4-job 3-machine problem (Figure 7(a)), as reported in our earlier work [4]. There are N = 30 variables and M = 48 constraints in the formulation of a 4-job 3-machine problem. The result turned out to be a near-optimum schedule as shown in Figure 7(b). One reason for not finding the optimum solution is probably due to the fact that the highly convoluted energy surface has many local minima.

6.6 DISCUSSIONS

We have shown how a modified Tank and Hopfield network can be utilized to solve difficult optimization problems such as the classical job-shop shceduling. Zhou et. al. [2] also presented a modified Tank and Hopfield analog computational network which exhibits linear scaling property, and thus appears to be better than the original approach by Foo and Takefuji [4]. However, a closer look reveals that their approach requires extensive computations by each neural processor. Thus, there is less network complexity at the expense of more complex processing by each neuron. This contradicts the popular concept of neurons with simple activation functions. Our approach may require more neurons and interconnects, but each neuron has a very simple activation function. Thus, our network is self-contained and does not need extensive calculations as required in the approach by Zhou et. al.

The software simulator written to simulate the behavior of the network uses the 4th-order Runge-Kutta integration technique to solve the differential equations (5) and

(6). So far, only small problems were simulated. Results shows that as the problem size increases, there is no guarantee of finding an optimum solution using this deterministic analog computational network but the network will always provide very good valid solutions.

6.7 ACKNOWLEDGEMENTS

The authors would like to thank Lisa R. Anderson and Joe Cordaro for implementing the linear programming circuit which was later modified to solve the job-shop scheduling.

6.8 REFERENCES

[1] D. W. Tank and J. J. Hopfield, "Simple 'Neural' Optimization Networks: An A/D Converter, Signal Decision Circuit, and a Linear Programming Circuit," IEEE Trans. on Circuits and Systems, Vol. CAS-33, No. 5, pp.533-541, May 1986.

[2] D. N. Zhou, V. Cherkassky, T. R. Baldwin, and D. E. Olson, "A Neural Network Approach to Job-Shop Scheduling," IEEE Transactions on Neural Networks, Vol. 2, No. 1, pp.175-179, January 1991.

[3] S. Y. Foo, L. R. Anderson, and Y. Takefuji, "Analog Components for the VLSI of Neural Networks," IEEE Circuits and Devices, Vol. 6, No. 4, pp.18-26, July 1990.

[4] S. Y. Foo and Y. Takefuji, "Integer-Linear Programming Neural Networks for Job-Shop Scheduling," Proc. of IEEE IJCNN '88, San Diego, CA, pp. 341-348, 1988.

[5] Y. Takefuji, <u>Neural Network Parallel Computing</u>, Kluwer Publishers, 1992.

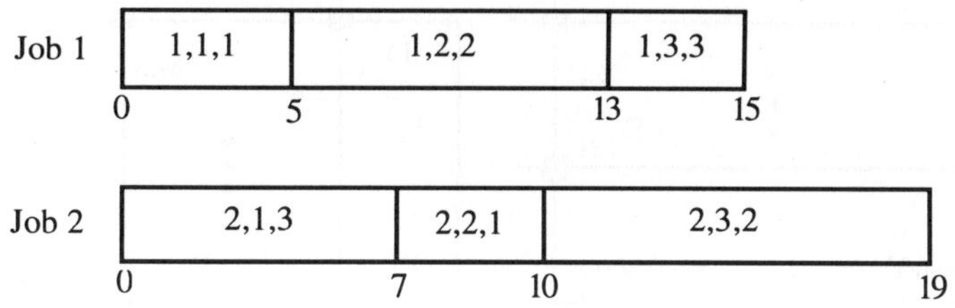

(a)

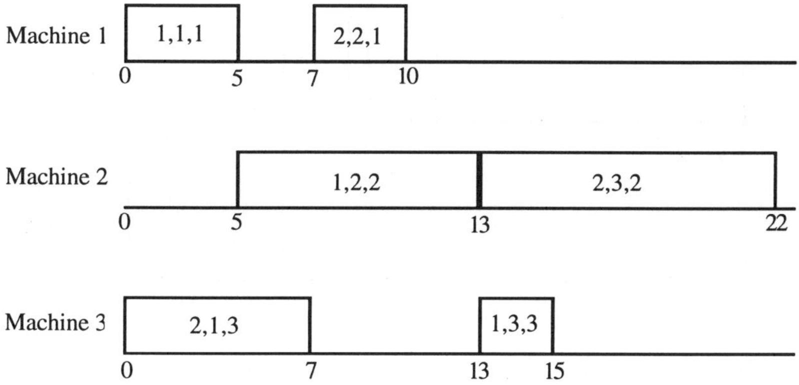

(b)

Fig. 1(a) An example 2-job 3-machine job-shop problem. (b) The optimum schedule as produced by the network.

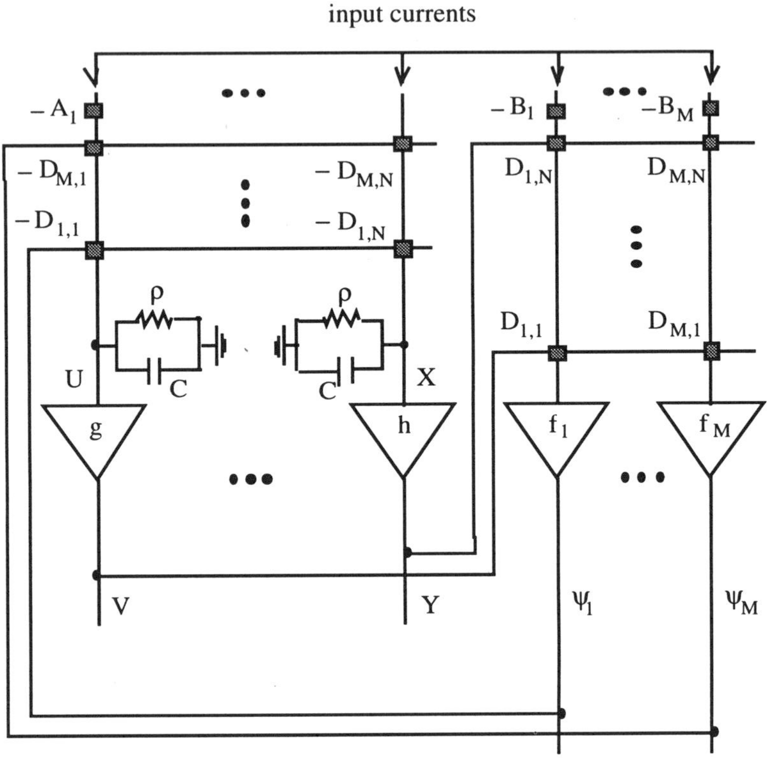

Fig. 2 The modified Tank and Hopfield network consisting of linear and nonlinear processors for solving mixed integer linear programming.

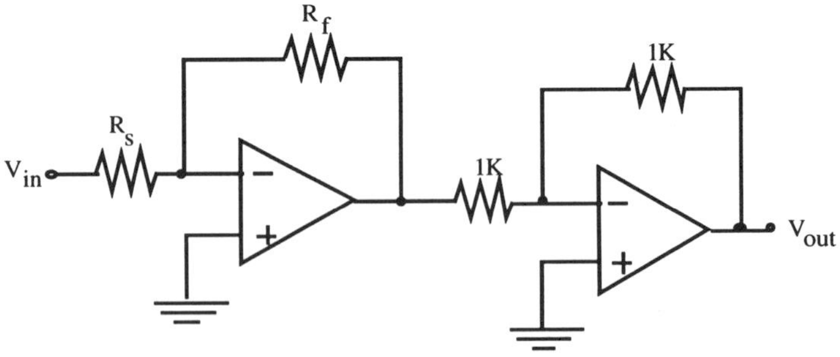

Fig. 3 A noninverting amplifier circuit for the linear g-amplifier.

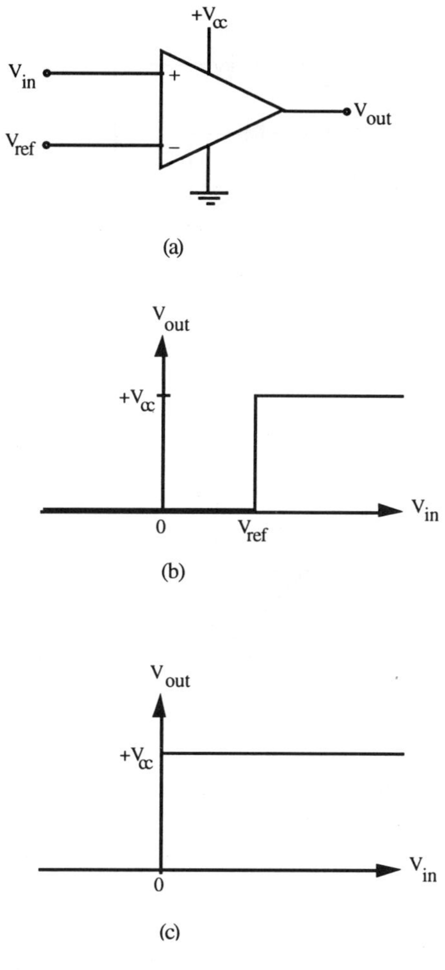

Fig. 4(a) An analog comparator for implementing an h-amplifier. (b) The adjusting step function. (c) The neutral position with $V_{ref} = 0$.

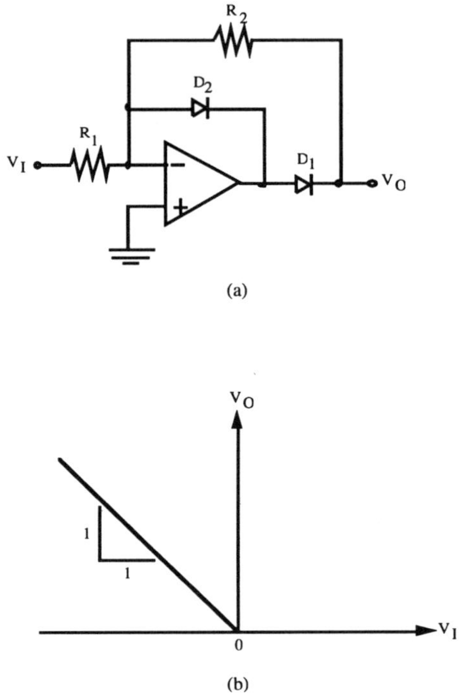

Fig. 5(a) A precision inverting half-wave rectifier implementing the f-amplifier circuit. The diode D_2 keeps the feedback path closed around the op-amp during the times when rectifier diode D_1 is in cut-off, thus preventing the op-amp from saturating which causes slower slew-rate. (b) The transfer function of the circuit for $R_1 = R_2$.

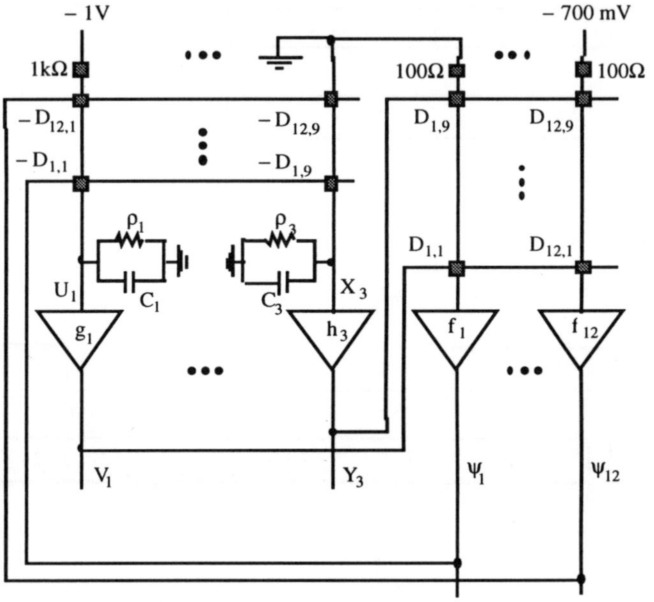

Fig. 6 Circuit diagram of the network for solving a 2-job 3-machine problem.

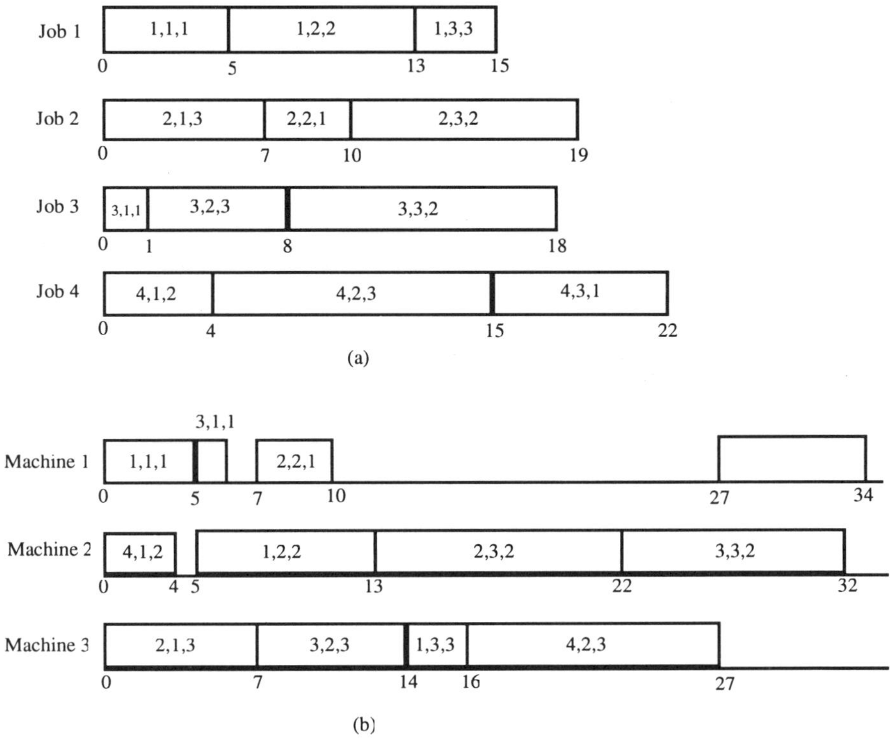

Fig. 7(a) An example 4-job 3-machine job-shop problem. (b) A near optimum schedule produced by the simulator.

Chapter 7

BIBD PROBLEMS

TAKAKAZU KUROKAWA
Department of Computer Science
The National Defense Academy
Yokosuka-shi 239, Japan

YOSHIYASU TAKEFUJI
Keio University, Japan
Case Western Reserve University, USA

Neural network parallel computing for balanced incomplete block design (BIBD) problems is presented in this section. A BIBD problem is a kind of block design problems consisting of collections of subsets of a set with certain balanced properties. These problems are very important for experimental design, material relating design, and coding theory. There are two representative methods to solve BIBD problems. One uses the notion of the finite fields, and the other uses the notion of the difference sets. But in general, there have been no parallel algorithms for BIBD problems. Furthermore, the algorithm using the notion of the finite fields is only able to solve the limited problems systematically when they satisfy an affine plane or a finite projective plane. Here, we will show a neural network which can not only solve BIBD problems where the conventional algorithms can solve using finite fields, but also can solve the other BIBD problems which will not satisfy the finite fields. These algorithms can be easily implemented by the analog circuits. The proposed algorithms were verified by a large number of simulation runs. These simulation results demonstrates an ability of the neural networks to search a solution out of a large space.

7.1 BIBD PROBLEMS

A design in which all the blocks contain the same number of elements, and all the elements appear in the same number of blocks, is called a block design. A block is said to be incomplete if it does not contain all elements. If a design is balanced, it is called as abalanced incomplete block design (BIBD)(1). A BOBD plays an important role in the experimental design, and the modern theory of statistical analysis(2). It can be further applied to material relating design and coding theory(3).

There is a famous mathematics problem called "Kirkman's schoolgirl problem", proposed by T.P.Kirkman in 1850(4). A school teacher is in the habit of taking her fifteen girls denoted by $V=\{1,2,\cdots,15\}$ for a daily walk, always arranging them three abreast walk which is called a block and is denoted by B_i where i equals to $1,2,\cdots,35$ in five rows. The problem is to arrange them so that for seven consecutive days no girl will walk more than once in the same row with any other girl. Fig 7.1 shows one of the solutions of this problem.

The following conditions must be satisfied for a collection $\{B_1,B_2,\cdots,B_{35}\}$ of subsets of V to be a BIBD problem, or (15,35,7,3,1)-design problem:(2)

(a) Each subset B_i $(i=1,2,\cdots,35)$ of V, which is called a "block", contains three elements.

(b) Each element $x \in V$, which is called "variety", is in seven of the subsets B_i $(i=1,2,\cdots,35)$. The number '7' in this case is called the "replication number" of the design.

(c) Every pair x, y of elements of V appear together in one of the subsets B_i ($i=1,2,\cdots,35$). The number '1' in this case is called the "covalency" for the design.

Sun.	(1,2,3)(4,8,12)(5,10,14)(6,11,13)(7,9,15)
Mon.	(1,4,5)(2,8,10)(3,13,15)(6,9,14)(7,11,12)
Tues.	(1,6,7)(2,9,11)(3,12,14)(4,10,15)(5,8,13)
Wed.	(1,8,9)(2,12,15)(3,5,6)(4,11,14)(7,10,13)
Thurs.	(1,10,11)(2,13,14)(3,4,7)(5,9,12)(6,8,15)
Fri.	(1,12,13)(2,4,6)(3,9,10)(5,11,15)(7,8,14)
Sat.	(1,14,15)(2,5,7)(3,8,11)(4,9,13)(6,10,12)

Fig. 7.1 A solution of the Kirkman's schoolgirl problem.

In general, the conditions for (v,b,r,k,λ)-design problem are given by:

(I) Each subset B_i ($i=1,2,\cdots,b$) of V contains k elements where k is a fixed constant.

(II) Each element $x\in V$ is in $r(\leq b)$ of the subsets $B_i(i=1,2,\cdots,b)$.

(III) Every pair x, y of elements of V appears together in $\lambda(\leq b)$ of the subsets B_i ($i=1,2,\cdots,b$).

Two relationships among the five parameters v, b, r, k, and λ are given by:

1) $v\cdot r=b\cdot k$

2) $\lambda\cdot(v-1)=r\cdot(k-1)$.

These relationships are presented as a theorem and also proved in reference (2). Using these relationships, a (v,b,r,k,λ)-design problem can be expressed as a $[k,\lambda;v]$-design problem in an abbreviation form which we follow in this book.

Owing to the research of Fisher and Yates, the experimental design has played an important role in the modern theory of statistical analysis(12)(13)(14). Bose presented several mathematical methods to construct block designs(15)(16)(17).

Grimaldi presented in reference (2) a condition under which [k,λ;v]-design problems could exist and how such designs could be solved using finite fields. These research results are shown in reference (18) preciously.

Through the consecutive research more than a half century by mathematicians, there have already been proposed two representative methods to solve BIBD problems. One uses the notion of the finite fields, and the other uses the notion of the difference sets. But in general, the former method can only solve a set of problems which must satisfy an affine plane or a finite projective plane in systematic way, and the latter method can only solve the BIBD problems through the exhaustive search and cannot solve them systematically. The systematic method using the affine plane AG(n,q) to solve the [q,1;q^n]-design requires the following condition: q is a power of a prime number and n is a positive integer. Another systematic method using the finite projective plane PG(n,q) to solve the [q+1,1;$\frac{q^{n+1}-1}{q-1}$]-design requires the same conditions. In both methods, points in planes accord with varieties, and lines accord with blocks of a BIBD problem. If k and v do not satisfy this relationship, these conventional methods cannot solve the [k,λ;v]-design problem although solutions may exist. For example, it is impossible to solve the [3,1;13]-design problem using the finite fields, because it does not satisfy the affine plane or the finite projective plane. Another solution method using the notion of the difference set can prove the existence of a solution, while it requires an exhaustive search to find out an initial set for the rotation sets(1). None have attempted yet to solve the BIBD problems in parallel.

7.2 NEURAL NETWORKS FOR BIBD PROBLEMS

7.2.1 NEURAL REPRESENTATION

Here, we will show a neural network representation to solve a [k,λ;v]-design problem. A neural network for a [k,λ;v]-design problem consists of $\binom{v}{k}$ binary neurons, each of which corresponds to a candidate block representing a possible combination of k elements out of v varieties. The binary neurons are what is called the McCulloch-Pitts neurons. The input and output of the ith binary neuron at time t are given as $U_i(t)$ and $V_i(t)$ respectively. The input/output function of the ith neuron is given by

$$V_i(t) = \begin{cases} 1, & \text{if } U_i(t) > 0 \\ 0, & \text{otherwise} \end{cases} \quad \text{for } i=1,2,\cdots,\binom{v}{k}. \tag{7.1}$$

Consider a simple example of the BIBD problems; a [3,1;7]-design problem. There are seven varieties which we denote 1,2,⋯,7 here. The number of possible 3-element candidates to make a block is calculated as $\binom{7}{3}$=35. The state of a single neuron described by a square in Fig 7.2 indicates the selection of the corresponding blocks. One of the solutions of a [3,1;7]-design problem is (1,2,3) (1,4,7) (1,5,6) (2,4,6) (2,5,7) (3,4,5) (3,6,7), and the selection of the required candidates is described by the black squares in Fig 7.2.

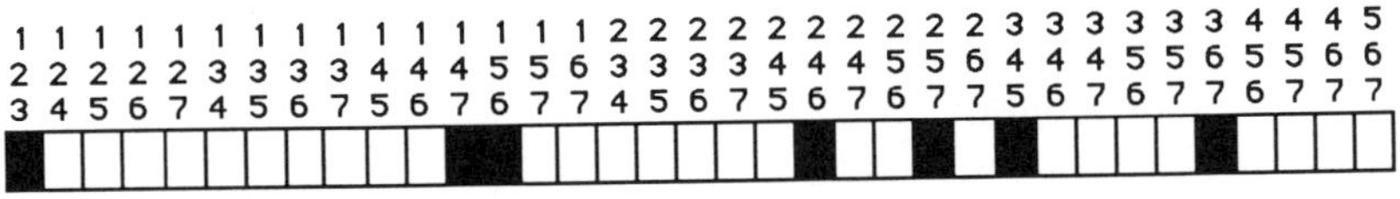

Fig. 7.2 Neural network representation for [3,1;7]-design problem.

A block diagram of the neural network for [3,1;7]-design problem is shown in Fig 7.3. In this figure, each neuron is represented as a shape of an inverter. The output of each neuron performs an inhibitory force to avoid the other candidates which have at least two same elements in their variety to be selected in a solution.

Therefore, an input of a neuron is connected to the outputs of those neurons which have at least two same elements in their variety as shown in Fig 7.3. For example, the input of the leftmost #1 neuron is connected to the outputs of #2,#3,···,#9,#16,#17,#18,#19 neurons, because all of these neurons have at least the same pair of elements (1,2), (1,3), or (2,3). The output state of the i-th neuron V_i for $i=1,2,\cdots,\binom{v}{k}$ becomes 1 if the corresponding i-th block is in the design and becomes 0 otherwise. The motion equation of the i-th neuron for $i=1,2,\cdots,\binom{v}{k}$ is given by:

$$\frac{dU_i}{dt} = -A\sum_{\substack{j=1\\ j\neq i}}^{\binom{v}{k}} f(N_j,N_i)\cdot V_j + Bh\left(\sum_{\substack{j=1\\ j\neq i}}^{\binom{v}{k}} f(N_j,N_i)\cdot V_j\right) \tag{7.2}$$

where A and B are both coefficients in integer region. The function f(x,y) is 1 if the block x and y have at least a pair of the same elements in their variety, and is 0 otherwise. This function is realized as connections of an input of a neuron to the outputs of the relating neurons in a real neural network. The function h(x) is a deterministic hill-climbing function where

$$h(x) = \begin{cases} 1 & \text{if } x=0, \\ 0 & \text{otherwise.} \end{cases} \tag{7.3}$$

The first term in Eq.2 is the inhibitory force in order to avoid any two blocks, which have at least a pair of the same elements to be selected in a solution. The second term is the encouragement force to select the i-th block if the combination of k elements applied to the i-th block causes no conflict to the other blocks which are already selected in a solution.

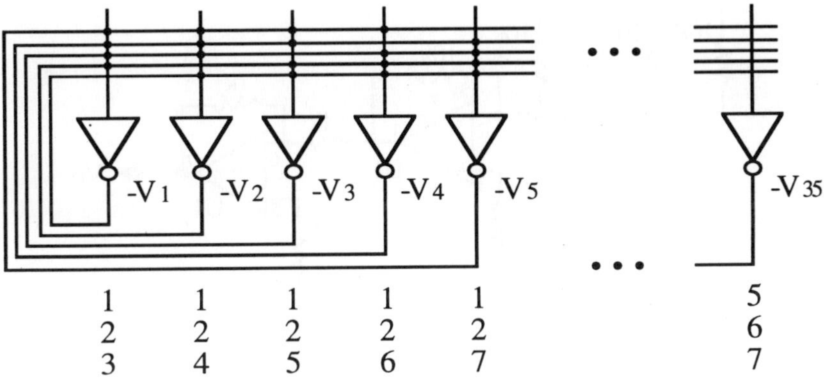

Fig. 7.3 A block diagram of a neural network for [3,1;7]-design problem.

7.2.2 ANALOG CIRCUITS

Fig 7.4 shows a complete circuit diagram of #1 neuron which have already shown in Fig 7.3 as a block diagram to solve [3,1;7]-design problem. The entire circuit is divided into four parts here.

Fig 7.4(a) shows the circuit diagram of the summing operation $\sum_{\substack{j=1 \\ j\neq 1}}^{35} f(N_j,N_1)\cdot V_j$. The function $f(N_j,N_1)$ is realized by a connection of the output V_j to the input of #1 neuron through a $10k\Omega$ resistor. The operational amplifier performs summation of the connected outputs V_j and determines the output voltage V_o as

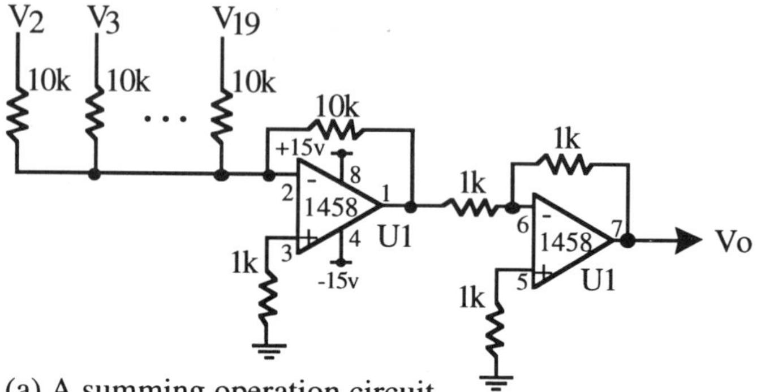

(a) A summing operation circuit.

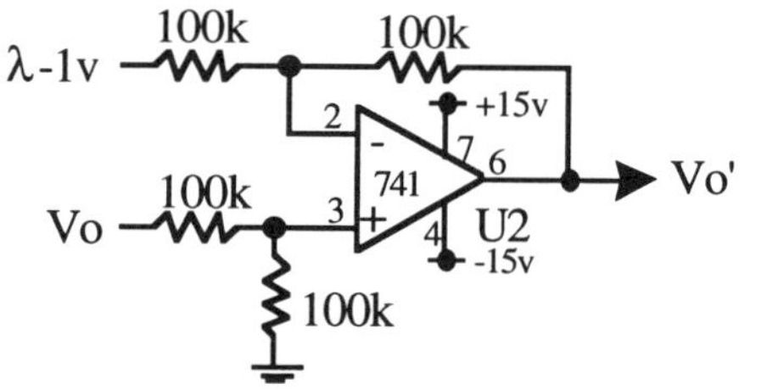

(b) A difference operation circuit.

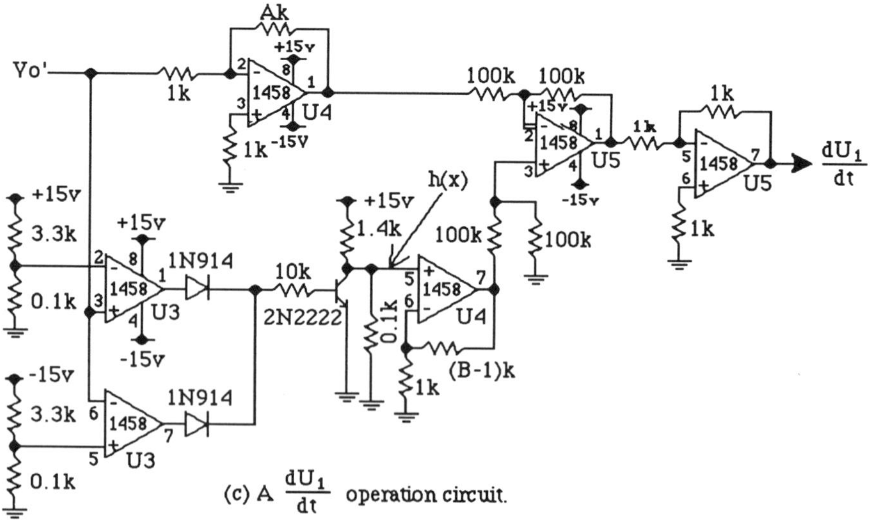

(c) A $\frac{dU_1}{dt}$ operation circuit.

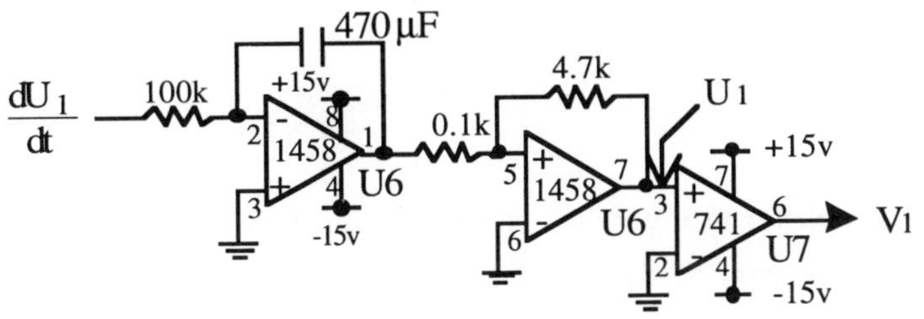

(d) An integrator and a V1 operation circuit.

Fig. 7.4 An analog circuit of the #1 neuron.

$$V_o=V_2+V_3+\cdots+V_{19}. \tag{7.4}$$

In Fig 7.4(b) the operational amplifier subtract the value $(\lambda\text{-}1)$ from V_o and generates the calculation result expressed as V_o'. In Fig 7.4(c) a circuit to calculates the value of $\frac{dU_1}{dt}$ discussed in Eq.2 is shown. The input of this circuit is V_o' derived from the difference operation circuit shown in Fig 7.4(b). An inverting amplifier shown in upper left part in Fig 7.4(c) multiplies its input by A. The lower left components consisting of two inverting amplifier, two diodes, a transistor, and seven resistors operates as the binary function determined by Eq.3. The output of this component is multiplied by B through a noninverting amplifier. Two outputs of an inverting and a noninverting amplifier are summed up by the rightmost two operational amplifiers. The output of this circuit generates $\frac{dU_1}{dt}$. In Fig.4(d), the first operational amplifire performs integration which is given by:

$$-\frac{1}{CR}\int\left(\frac{dU_1}{dt}\right)dt = -\frac{U_1}{CR} = -\frac{U_1}{47}\,. \tag{7.5}$$

The second operational amplifire generates U_1:

$$-\left(-\frac{U_1}{CR}\right)R' = U_1 \,. \tag{7.6}$$

The last operational amplifier in Fig 7.4(d) is a binary function determined by the Eq.1. The other neurons from #2 to #35 can also be designed in the same way shown above. The only difference among 35 neurons is the connection of their inputs and outputs.

7.2.3 PARALLEL ALGORITHM

The following procedure describes a parallel algorithm to simulate the neural network for BIBD problems.

[Step 0] Set t=0.

[Step 1] The states of inputs $U_i(t)$ for $i=1,2,\cdots,\binom{V}{k}$ are set to the small negative number or randomized.

[Step 2] Evaluate the output values $V_i(t)$ for $i=1,2,\cdots,\binom{V}{k}$ based on the binary function in Eq.1.

[Step 3] Use the motion equation in Eq.(2) to compute $\Delta U_i(t)$.

$$\Delta Ui(t)=-A\sum_{\substack{j=1\\ j\neq i}}^{\binom{v}{k}} f(N_j,N_i)\cdot V_j+Bh\left(\sum_{\substack{j=1\\ j\neq i}}^{\binom{v}{k}} f(N_j,N_i)\cdot V_j\right). \tag{7.7}$$

[Step 4] Compute $U_i(t+1)$ on the basis of the first order Euler method:

$$U_i(t+1)=U_i(t)+\Delta U_i(t) \quad \text{for } i=1,2,\cdots,\binom{V}{k}. \tag{7.8}$$

[Step 5] Increment t by 1.

If the number of fired neuron equals to $b(=\lambda v(v-1)/k(k-1))$ and if the generated element pairs are all unique, then terminate this procedure, else go to step 2.

7.2.4 SIMULATION RESULTS

Following the parallel algorithm presented above, neural network simulators for BIBD problems have been developed on a Macintosh SE/30, Sun Sparc station, DEC station, and NeXT station using Pascal or C language.

The relationships between the frequency and the number of iteration steps to converge to the global minimum for [3,1;7], [3,1;9], and [4,1;13]-design problems through a thousand simulation runs are shown in Fig 7.5, Fig 7.6, and Fig 7.7 respectively. The number of possible candidates, the complexity of the searching space for these three problems are $\binom{35}{7}=6.7\times10^6$, $\binom{84}{12}=1.1\times10^{14}$, and $\binom{715}{13}=1.8\times10^{27}$ respectively. The simulation result depicts that the state of the system converged to the solution within several tens iteration steps. The following parameters were used in our simulations: A=1, B=k-1 if the number of blocks selected in the design is greater than 2/3 of b, otherwise B=k-2.

Our algorithm solved the [3,1;15]-design problem in 159 iteration steps, which is the famous Kirkman's schoolgirl problem shown in n.1. The number of possible candidates, the complexity of the searching space is $\binom{455}{35}=2.7\times10^{52}$. Note that the mass of a mountain is about 10^{43} times that of an electron(19).

Furthermore the [3,1;13]-design problem, which cannot be solved by the conventional algorithms using finite fields, were solved by our simulator. Fig 7.8 shows the simulation result of this problem. In Fig 7.8, the state of 286 neurons is given by a 11*26 rectangular array, where the black squares indicate the selection of blocks. In the triangular array, the coordinate (i,j) for i=2,3,...,13, and j=1,2,...,12 indicates whether a pair of variety i and j exists ("1") or not ("0"). One of the solutions of this problem is (1,2,6) (1,3,9) (1,4,5) (1,7,13) (1,8,10) (1,11,12)

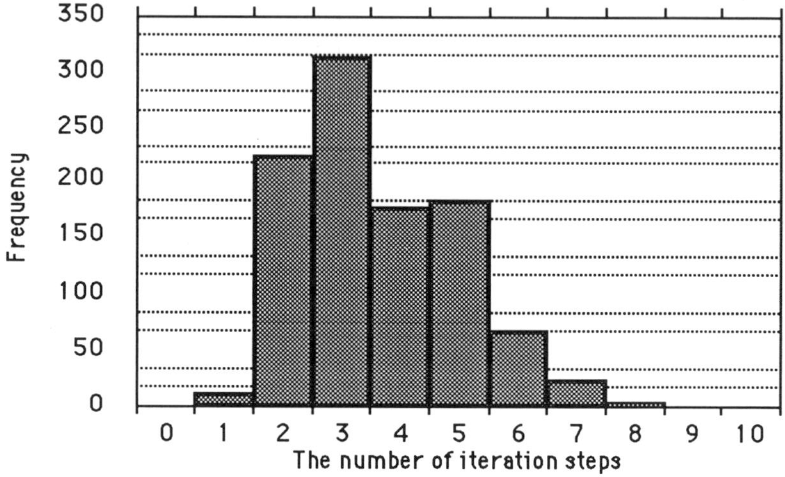

Fig. 7.5 The relationship between the frequency and the number of iteration steps to conberge to global minimum in the [3,1;7]-design problem.

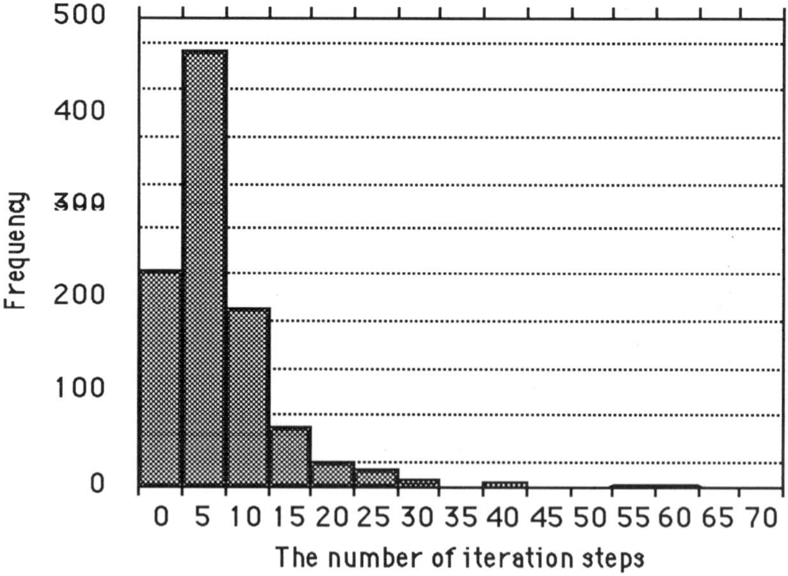

Fig. 7.6 The relationship between the frequency and the number of iteration steps to converge to global minimum in the [3,1;9]-design problem.

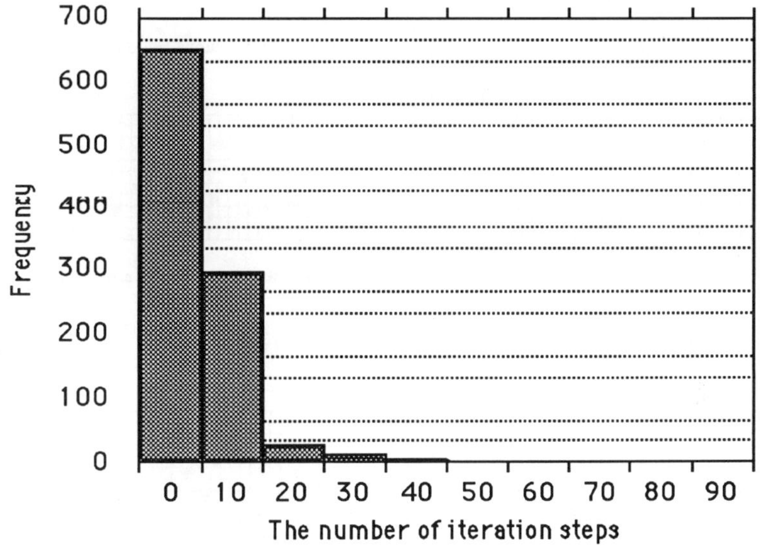

Fig. 7.7 The relationship between the frequency and the number of iteration steps to converge to global minimum in the [4,1;13]-design problem.

(2,3,11) (2,4,10) (2,5,7) (2,8,12) (2,9,13) (3,4,13) (3,5,8) (3,6,12) (3,7,10)(4,6,8) (4,7,12) (4,9,11) (5,6,9) (5,10,11) (5,12,13) (6,7,11) (6,10,13) (7,8,9)(8,11,13) (9,10,12) where the variety is numbered from 1 to 13. The complexity of the searching space is $\binom{286}{26}=5.6\times10^{36}$. Note that the mass of a man is about 10^{31} times that of an electron(19).

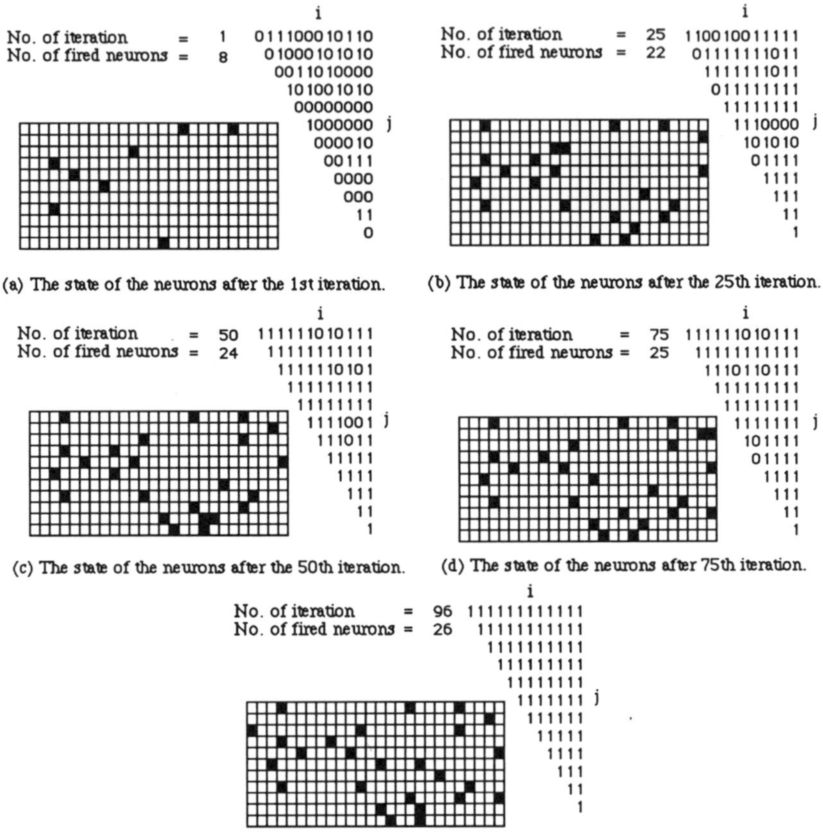

(a) The state of the neurons after the 1st iteration.

(b) The state of the neurons after the 25th iteration.

(c) The state of the neurons after the 50th iteration.

(d) The state of the neurons after 75th iteration.

(e) The state of the neurons of the final stage.

Fig. 7.8 The simulation result of a [3,1;13]-design problem.

REFERENCES

(1)A.P.Street, and D.J.Street, Combinatorics of Experimental Design (Clarendon Press, Oxford, 1987), p.400.

(2)R.P.Grimaldi, Discrete and Combinatorial Mathematics (Addison-Wesley Publishing Company, 1985).

(3)A.P.Street, and W.D.Wallis, Combinatorial Theory: An Introduction (The Charles Babbage Research Center, Winnipeg, Canada, 1977).
(4)M.Gardner, New Mathematical Diversions from Scientific American (University of Chicago Press, Chicago, 1966).
(5)J.Hopfield, and D.Tank, Biological Cybernetics, **52** (1985) 141-152.
(6)Y.Takefuji, and K.C.Lee, Science, **245** (1989) 1221-1223.
(7)Y.Takefuji, and K.C.Lee, IEEE Trans. on Neural Networks, **1** (1990) 143-145.
(8)Y.Takefuji, and K.C.Lee, J. Neural Network Computing, (1990).
(9)Y.Takefuji, and K.C.Lee, IEEE Trans. on Circuits and Systems, **37** (1990) 1425-1429.
(10)Y.Takefuji, C.W.Lin, and K.C.Lee, Biological Cybernetics, **63** (1990) 337-340.
(11)Y.Takefuji, L.L.Chen, K.C.Lee, and J.Huffman, IEEE Trans. on Neural Networks, **1** (1990) 263-267.
(12)R.A.Fisher, J. Minis. Agric. **33** (1926) 503-513.
(13)R.A.Fisher, and F.Yates, Agricultural and Medical Research, Oliver and Boyd (1938).
(14)R.A.Fisher, Ann. Eugenics, **10** (1940) 52-75.
(15)R.C.Bose, Ann. Eugenics, **9** (1939) 353-399.
(16)R.C.Bose, Bull. Calcutta Math. Soc., **34** (1942) 17-31.
(17)R.C.Bose, and S.S.Shrikhande, Canad. J. Math., **12** (1960) 177-188.
(18)I.Anderson, Combinatorial Designs: Construction Methods, Ellis Horwood Limited, (1990).
(19)R.K.Adair, The great design: particles, fields, and creation, Oxford University Press, (1989).

Chapter 8

DISCOVERING RNA INTERACTIONS

YOSHIYASU TAKEFUJI
Dept. of Electrical Engineering and Applied Physics
Case Western Reserve University
Cleveland, OH 44106, USA
and
Faculty of Environmental Information
Keio University, Fujisawa 252, Japan

DORA BEN-ALON
ARIEH ZARITSKY
Department of Life Sciences, Ben-Gurion University of the Negev,
P.O. Box 653, Be'er-Sheva 84105, Israel

High-order RNA structures are involved in regulating many biological processes; various algorithms have been designed to predict them. Experimental methods to probe such structures and to decipher the results are tedious. Artificial Intelligence and the Neural Network approach can support the process of discovering RNA structures.

Secondary structures of RNA molecules are probed by autoradiographing gels, separating end-labeled fragments generated by base-specific RNases. This process is performed in both conditions, denaturing (for sequencing purposes) and native.

The resultant autoradiograms are scanned using line-detection techniques to identify the fragments by comparing the lines with those obtained by 'alkaline ladder's. The identified paired bases are treated by either one of two methods to find the foldings which are consistent with the RNases' 'cutting' rules. One exploits the maximum independent set algorithm, the other - the planarization algorithm. They require, respectively, n and n2 processing elements, where n is the number of base pairs. The state of the system usually converges to the near-optimum solution within about 500 iteration steps, where each processing element implements the McCulloch-Pitts binary neuron. Our simulator based on the proposed algorithm discovered a new structure in a sequence of 38 bases, which is more stable than the formerly proposed.

8.1 INTRODUCTION

The four common bases of an RNA molecule are cytosine (C), uracil (U), adenine (A) and guanine (G). The double-helix of an RNA forms when two separate sections (with a 5'-end to 3'-end polarity) become linked together in an anti-parallel manner by weak hydrogen bonds between specific, complementary bases: A always pairs with U and G pairs with C. Primary structure of RNA is defined as its linear base sequence. Secondary structure is determined by its folding into a two-dimensional shape. Folding into a three-dimensional shape is called tertiary structure, and structures formed by interactions with other molecules are quaternary.

RNA molecules are involved in a wide range of functions in the living world, exerted in part by the three-dimensional conformations to which they can fold (e.g., Zaritsky et al. (1988); Dahlberg & Abelson (1989, 1990); Puglisi et al., (1991)). The stability of a structure is measured by the free-energy difference between folded and unfolded forms. RNA sequence can often form alternate structures of similar stabilities, which may be the reason for its role in various processes (e.g., translation). Predictions of the secondary structure of RNA, that is, its base-pairing pattern (whether based on free energy calculations or distance geometries or inferred from compensatory mutations; e.g., Zuker & Stiegler (1981); Williams & Tinoco (1986); Martinez (1988); Zuker (1989); Mei et al. (1989); Takefuji et al. (1990a and 1990b); Major et al. (1991)) are more reliable than those predicted for proteins from amino acid sequences (Karplus & Petsko (1990)).

8.1.1 CURRENT APPROACHES TO THE PROBLEM

Fresco (Fresco et al. (1960)) used the first model to predict secondary structures in RNA. Two types of algorithms have been reported: the combinatorial method (Pipas & McMahon (1975)) and the recursive (or dynamic programming) method (Nussinov et al. (1978)). Both algorithms, as well as the latest method proposed by Zuker (1989b), are based on sequential computation. Unfortunately, few parallel algorithms based on molecular thermodynamics models have been reported. Recently, Qian & Sejnowski (1988), and Holley and Karplus (1989) have reported a backpropagation algorithm using a three-layer-feed-forward neural network for protein secondary structure prediction. Their method is based on the correlation between secondary structure and amino acid sequences, but has the following drawbacks as compared with the conventional RNA folding algorithms based on molecular thermodynamics models: 1. They need a teacher to force the network to learn the correlation between a secondary structure and an amino acid sequence. 2. They cannot provide accurate prediction if a completely uncorrelated new datum is given where the previously learned correlation information is useless. 3. Their feed-forward neural network requires a prohibitively long learning process to deal with a long sequence of bases for RNA secondary structure prediction. 4. No theorem is given to determine the neural network architecture including how many hidden layers and how many hidden neurons per hidden layer should be used. Our algorithms (Takefuji et al. (1990a and b); and see Section 2) requires neither a teacher nor a learning process. The proposed maximum independent set parallel algorithm can yield the suboptimum solution within several hundred iteration steps using n processors (where n is the number of possible base pairs).

Generally speaking, the existing algorithms can be classified into three quite different approaches. One approach is phylogenetic structure analysis of homologous RNAs that depends on multiple alignment of the molecules (Jaeger et al. (1989); Le & Zuker (1991)). Helix conservation is scored by a ratio of the number of times that the helix occurs in suboptimal foldings. The combination of compatible helices generates a secondary structure of the statistically more significant ones.

The property of RNA molecules to fold has yielded algorithms of the second approach, that compute optimal foldings with mathematical tools, based on either maximizing number of pairings or minimizing free-energy. Recursive algorithms have been used and a collection of secondary structures that can be found close to the energy minimum are generated (Le & Zuker (1991)). Using Turner et al.'s (1987) energy rules, the computer prediction accuracy was elevated to 70%. Predictions made within 10% of the lowest free energy includes in them up to 90% of the phylogenetically known helices. Of course, the main problem remains to be solved, i.e., choosing the correct structure occurring in any given natural RNA. In addition, algorithms of this type require computer time proportional to nm, where n is the number of variables and m the number of permitted values.

A third, new approach, that of neural representation, encodes the problem using artificial neurons (Steeg (1989); Takefuji et al. (1990a; 1990b)). A fired neuron represents a possible base pair like G-C and A-U. This method resides in the energy family, and the idea is to find the largest number of base pairs which will prove to have the minimum energy. Use of graph theory solution for finding the largest planar subgraph or the maximum independent set permit finding the largest set base pairs. Among the diverse structures, the more knowledges and rules from experts embedded in the neural network, the better is the solution and the faster it is obtained.

These models, albeit theoretical, do consider actual information available about chemical interactions between bases. For example, the nearest neighbor model approximates the stability of an RNA duplex by the sum of the free energy increments for all its ten nearest neighbors in the duplex provided by various measurements (e.g., Freier et al. (1986); Turner et al. (1987); Jaeger et al. (1989)). However, the algorithms based on free energy minimization or distance geometry are limited because solutions might represent local minima, depending on the input structure, rather than the global minimum. There can therefore be numerous foldings within 5 to 10% of the computed minimum free energy.

Uncertainties and difficulties of these kinds can be mitigated by incorporation of additional data. For instance (Zuker (1989b)), incorporating nuclease data which identify single or double stranded regions results in a dot plot with compatible base pairs only. Further information is gained by determination of a common RNA secondary structure within a set of homologous RNAs (Jaeger et al. (1989); Le & Zuker (1991)) and analysis of time intervals in structural reconstructions (because both the building of an mRNA molecule and its passage into the cytoplasm of eukaryotic cells start from its 5' end; Gultyaev (1991)). All of these procedures are not sufficient to yield the correct, bio-active structure occurring for any given RNA; the main problem is thus not yet resolved. Some interactions between unpaired bases of a folded RNA and between RNA or other regulatory molecules in a living cell complicate the achievement of meaningful conclusions still further (Garrett et al. (1981); Goringer & Wagner (1988)).

8.2 MAXIMUM INDEPENDENT SET (MIS) PROGRAMS

8.2.1 THE ALGORITHM

An independent set in a graph is a set of vertices, no two of which are adjacent. A maximum independent set (MIS) is an independent set whose cardinality is the largest among all independent sets. The problem of finding an MIS for arbitrary graphs is nondeterministic polynomial-complete. The MIS problem, the max cut problem, and the maximum clique problem are all interrelated with each other for finding ground states of spin glasses with exterior magnetic field and solving circuit layout design problems in VLSI circuits and printed circuit boards.

Our parallel algorithm (Takefuji et al. (1990b)) generates a near-MIS of a circle graph in nearly constant time. For an n-edge problem, the algorithm uses n processing elements, each implementing the McCulloch-Pitts (1943) neuron model.

Consider the simple circle graph (Fig. 1a) with 14 vertices and 7 edges. Fig. 1b shows its adjacency graph, generated by edge-intersection in the circle graph. For example, the edge "d" intersects with three edges: c, e, and f. The adjacency graph G(V, E) of the circle graph is given by V={a, b, c, d, e, f, g} and E={(a b), (b c), (b e), (b f), (c d), (d e), (d f), (e f), (f g)}. A set of vertices {a, c, e, g} in Fig. 1c is the MIS of this circle graph which is equivalent to the maximal planar subgraph, as shown in Fig. 1d. In other words, finding the MIS in a circle graph is equivalent to finding its maximum planar subgraph. In order to find the near-maximal planar subgraph in the circle graph with m-vertex and n-edge, n neurons (processing elements) are used in our algorithm. The output state of the ith neuron Vi=1 means that the ith edge is not embedded in the circle graph. The state of Vi=0 indicates that the ith edge is embedded in the circle graph.

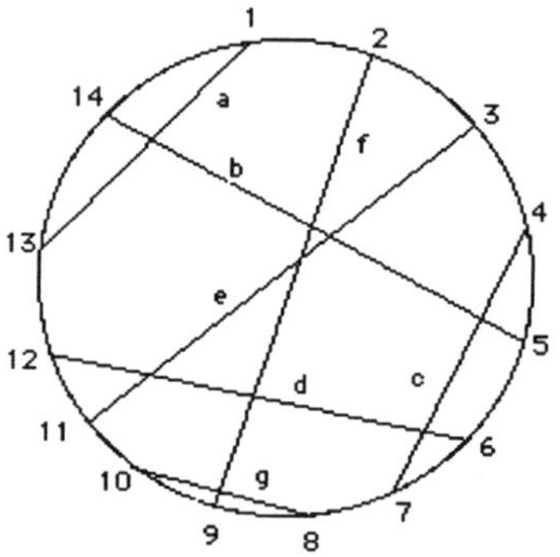

Fig. 1(a) Circle graph with 14 vertices and 7 edges.

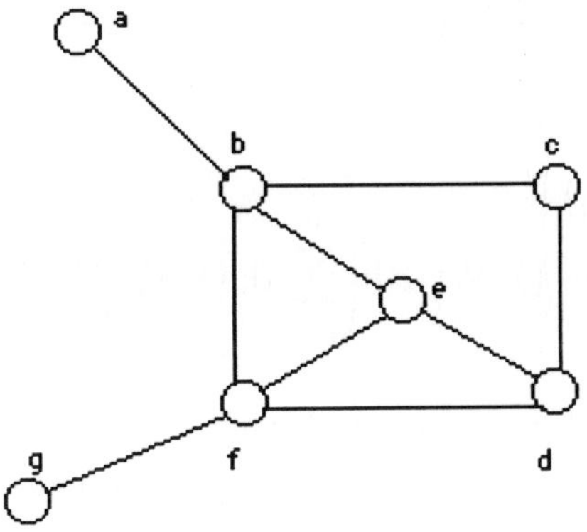

Fig. 1(b) Adjacency graph of Fig. 1a.

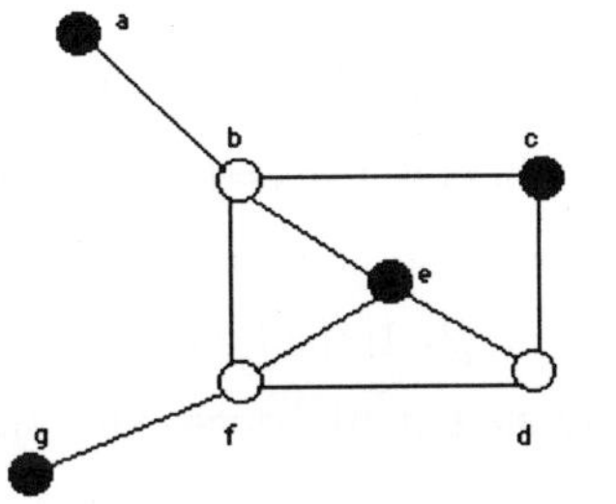

Fig. 1(c) Maximum independent set of Fig. 1a.

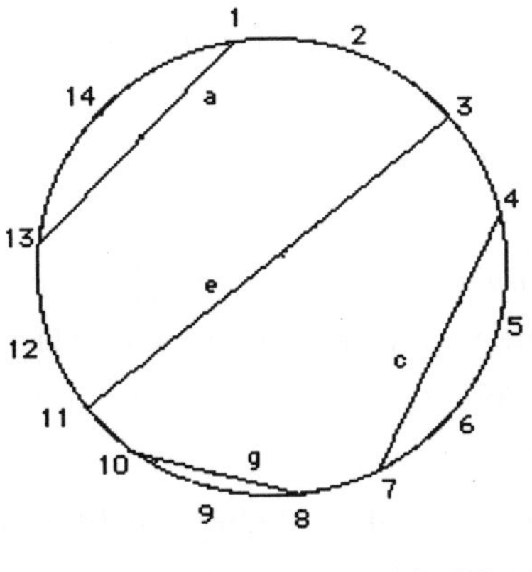

Fig. 1(d) Maximum planar subgraph of Fig. 1a.

The motion equation of the ith neuron for i=1,...,n is given by:

$$\frac{dU_i}{dt} = A\left(\frac{\sum_{j=1}^{n} d_{ij}(1-V_j)}{\text{distance(i)}}\right)(1-V_i) - Bh\left(\sum_{j=1}^{n} d_{ij}(1-V_j)\right)V_i \tag{1}$$

where dxy=1 if the xth edge and the yth edge intersect each other in the circle graph, 0 otherwise. Note that A and B are constant coefficients. Edge-intersection conditions between the ith and the jth edges in the circle graph are given by: head(i)<head(j)<tail(i)<tail(j) and head(j)<head(i)<tail(j)<tail(i) where tail(i) and head(i) are two end vertices of the ith edge. Note that distance(i) is given by distance(i)=min(|head(i) -tail(i)|, |n+head(i) -tail(i)|) where tail(i)>head(i) is always satisfied. The function h(x) is 1 if x=0, 0 otherwise.

The first term is the inhibitory force in order to remove the edges which intersect with the ith edge in the circle graph. If the ith edge is removed from the circle graph, the first term will not be activated at all, because the state of the ith neuron Vi=1. In order to keep the ith edge in the circle graph, the first term should not have any edge-intersection violation. Whenever the ith edge has any edge-intersection violation, it will tend to be eventually removed from the circle graph. The last term is the encouragement force to embed the ith edge in the circle graph. If the ith edge is removed but does not intersect with any other edges, the last term will force the ith neuron to be Vi=0. In other words, the ith edge is encouraged to exist in the circle graph.

Our goal is to maximize the number of edges in the planar circle graph where an edge represents a possible base-pair (G-C or A-U). The goal of maximizing the number of base pairs for predicting the secondary structure in RNA viroids was supported by Diener (1987).

The motion equation of Eq. 1 is slightly modified for predicting secondary structures of RNAs: (a) edge-intersection violation conditions must be updated. Six conditions to describe the edge-intersection in the circle graph are required: head(i)<head(j)<tail(i)<tail(j), head(j)<head(i)<tail(j)<tail(i), tail(i)=tail(j), tail(i)=head(j), head(i)=head(j), and head(i)=tail(j). The last four violation conditions are newly added to the first two, because a single base cannot be involved in more than one base-pair. The other modification is in the distance(i) function of Eq. 1, where it is given by distance(i)=|head(i) -tail(i)|.

A sequence of m bases is given to the simulator. It generates the circle graph with m vertices and n edges where n is the number of possible base pairs. Each base-pair must also satisfy the hairpin-loop constraint, |head(i) -tail(i)|>3, because it is sterically impossible to organize the hairpin loop with less than three bases. The circle graph is fed to the neural network simulator in order to find the near-MIS.

8.2.2 AN EXAMPLE

The stability number for a given RNA secondary structure is the sum of the contributions of the loops, bulges, and helices. The structure with the highest number is the most stable, called optimal folding. A sequence of 38 bases from residues 1118-1155 of Escherichia coli 16S rRNA served as an example to validate our simulator. Fig. 2a shows the secondary structure proposed by Stern (Stern et al. (1988)), where the structure stability (+7) is computed based on Tinoco's values: I. A-U pair, +1; II. G-C pair, +2; III. G-U pair, 0; IV. hairpin loops, -5 to -7; V. interior loops, -4 to -7; VI. bulges, -2 to -6.

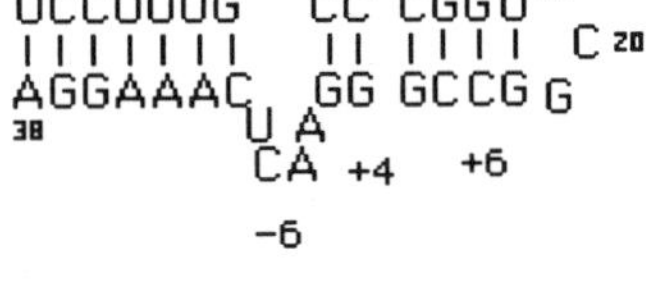

Fig. 2(a) Secondary structure of a sequence of 38 bases, proposed by Stern (Stern et al. (1988)).

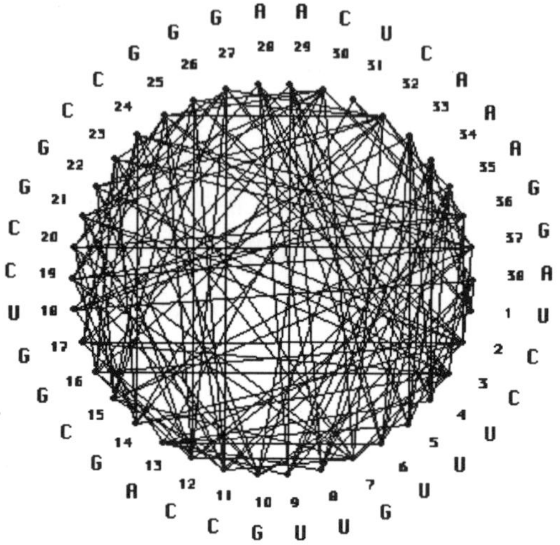

Fig. 2(b) Circle graph with 38 vertices and 151 edges.

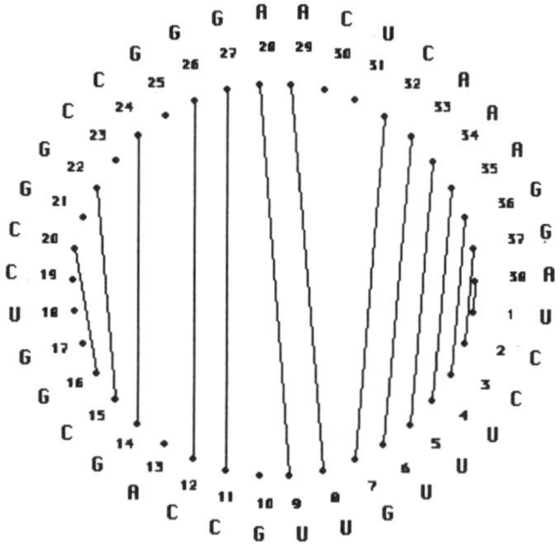

Fig. 2(c) State of the system after 104 iterations.

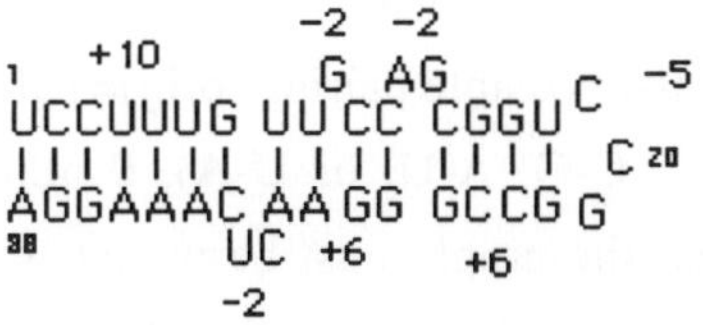

Fig. 2(d) Secondary structure predicted by our algorithm (Takefuji et al. (1990b)).

Fig. 2b shows the circle graph with 38 vertices and 151 edges,each edge represents a possible base-pairing.

When A=B=1 and Ui(0)=-5 for i=1,..., 151, the state of the system converged to the solution containing 14 edges (Fig. 2c) in the 104th iteration step. The secondary structure of the simulation result is given in Fig. 2d. Its stability number is +11, demonstrating that the simulator can find more stable structure than found by other means (Stern et al. (1988)).

8.3 PLANARIZATION AND RNA STRUCTURE PREDICTION

8.3.1 THE ALGORITHM

The mathematical problem to compute an optimal folding based on free-energy minimization is mapped onto a graph planarization problem (Takefuji et al. (1990b)). We want to maximize the number of edges in a plane with no two edges crossing each other. The A-U or G-C base pairs are only considered as possible edges to be embedded in a plane while the bases are the vertices. In other words, for a fragment stretching from ribonucleotides i to j, it is denoted by the subscript ijth neuron where the output and the input is depicted by Vij and Uij respectively for i=1,...,n-1 and j=i+1,...,n.

Consider a sequence of fifteen bases (Fig. 3a). In our algorithm a single-row representation is used where five edges are embedded. A (15 x 14)/2 neural network array (Fig. 3b) is used to predict the secondary structure of this problem. The following five functions are considered (Eq. 2): b(i,j), g(i,j,k), f(i,j), p(i,j,t), and h(x). The function b(i,j) denotes the possible pairing: b(i,j)=1 if i and j are one of the four legitimate base pairs (G-C, C-G, A-U, or U-A), 0 otherwise. Cross bonding is sterically impossible so that the graph must be planar without two edges crossing each other. A violation function g(i,j,k)=1 if i<j<k, 0 otherwise has been described (Takefuji & Lee (1989)). The function f(k,l) indicates the strength of a base pair bond between k and l bases: f(k,l)=2 if k and l bases are a G-C pair, 1 if they are an A-U pair, 0 otherwise (Tinoco et al. 1971). The hairpin loop constraint is also considered in our algorithm where more than two bases are required to make a hairpin loop: The function p(i,j,t)=1 if (j-hairpin)<i, 0 otherwise where hairpin is given by hairpin=4 if t=0, 55-t if 55-t >4, 4 otherwise. h(x) is the hill-climbing function, 1 if x=0, 0 otherwise.

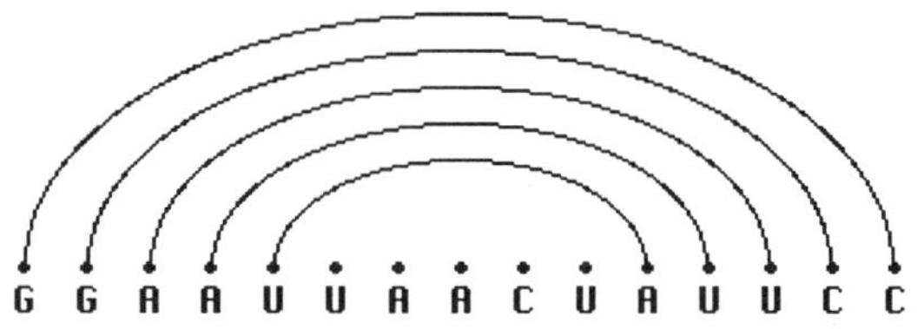

Fig. 3(a) Single-row representation of fifteen bases.

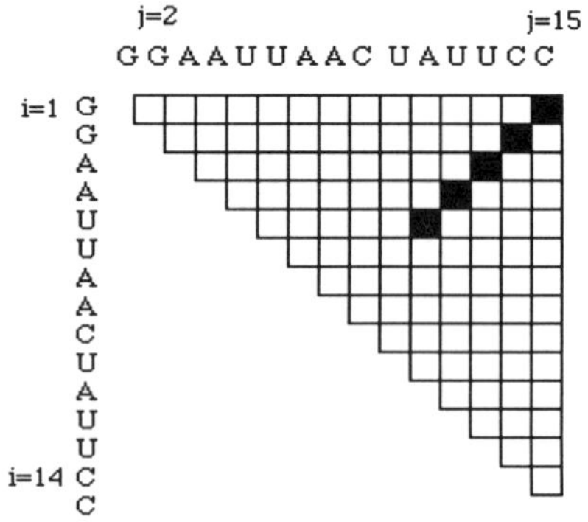

Fig. 3(b) Neural representation of a sequence of fifteen bases (Fig. 3a).

To predict suboptimal foldings of a sequence of n bases, n(n-1)/2 neurons are required. The motion equation of the ijth neuron is given by:

$$\frac{dU_{ij}}{dt} = -A_1\left(\sum_{k \neq i} V_{ik} - 1\right)b(i,j) - A_2\left(\sum_{k \neq j} V_{jk} - 1\right)b(i,j)$$
$$-B_1 \sum_{k<i<l<j} V_{kl}\, g(k,i,l)g(i,l,$$
$$-B_2 \sum_{i<k<j<l} V_{kl}\, g(i,k,j)g(k,j,l)f(k,l) \quad -Cp(i,j,t) \quad +Dh\left(\sum_{k \neq i} V_{ik}\right) \qquad (2)$$

The first two terms force the ith and jth bases, respectively, to have one and only one bond. Note that if the ith base has strong violations caused by other bases, it cannot have any bond. The third and fourth terms are inhibitory and always satisfy planarization conditions. The fifth term is the inhibitory hairpin constraint which prohibits less than three bases to make a hairpin loop. The hill-climbing force term (h(x)) allows the state of the system to escape from the local minimum.

8.3.2 AN EXAMPLE

Consider a sequence of 55 bases from R17 viral RNA (Tinoco et al.)1971)). 1485 (=55X54/2) neurons are used to solve this problem. Fig. 4, a and b, shows the state of the system after the 1st and the 61st iterations, respectively, the latter is translated to the the predicted secondary structure in Fig. 4c. The total stability of the structure is +7, which is equivalent to free energy ΔG=-8.4kcal per mol. When pairs (A15 to U41), (A16 to U40) and (A17 to U39) are shifted to (A15 to U40), (A16 to U39) and (A17 to U38), respectively, the total stability of the structure becomes +8, which is the optimum (Tinoco et al. (1971)).

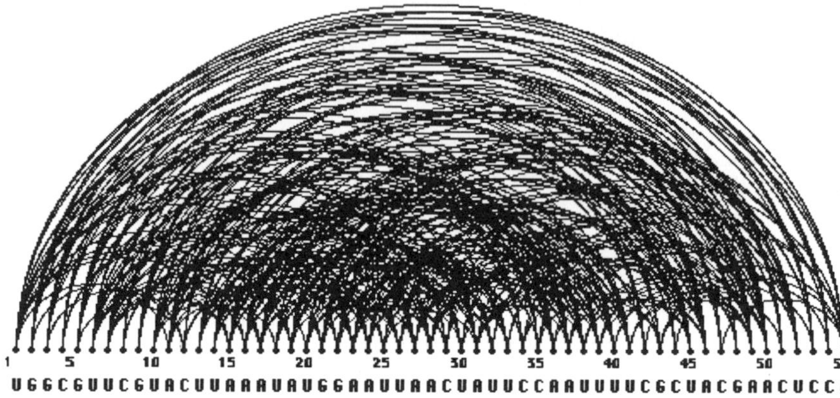

Fig. 4(a) State of the system after the first iteration step.

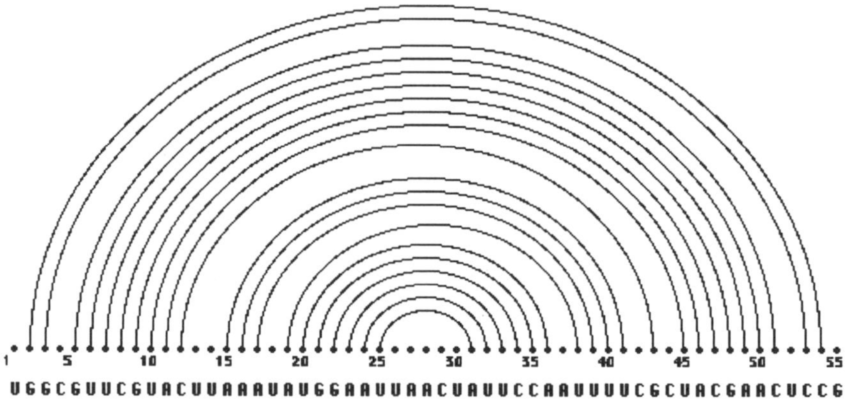

Fig. 4(b) State of the system after the sixty-first iteration step.

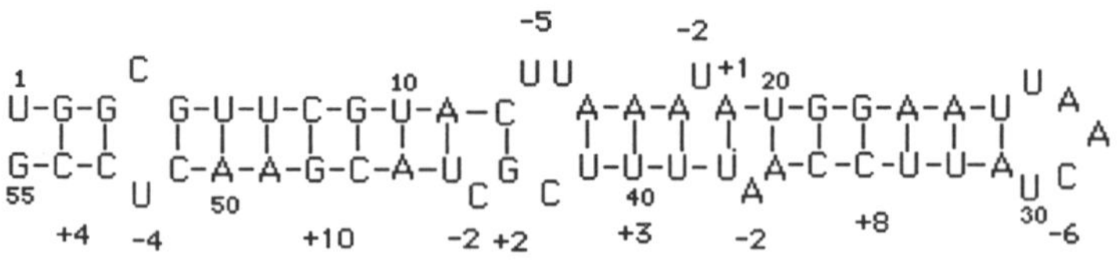

Fig. 4(c) Secondary structure predicted by our algorithm.

8.4 REAL LIFE ANALYSES

8.4.1 EXPERIMENTAL (WET) PROCEDURES

There exist several physico-chemical experimental procedures (e.g., Noller (1984)) to derive an RNA secondary structure, or to choose among alternative possibilities obtained by the conventional computer methods (Section 1.1). X-ray diffraction analysis (Holbrook et al. (1978)), electron-microscopy (Jacobson et al. (1985)) and absorption spectrum analysis (Reid (1981)), are some examples of direct methods, each with its own disadvantages. In all the indirect methods, the RNA molecule to be studied must be exposed to some chemical or enzymatic treatment and the nature and composition of products are analysed. These include: use of psoralens to cross-link the two DNA strands (e.g., Cimio et al. (1985)), binding of synthetic oligonucleotides (e.g., Mankin et al. (1981)) and the use of specific chemicals or specific ribonucleases (e.g., Ehresmann et al. (1987)).

The only way to tell which of the candidate structures really exists is by probing it chemically. Strong evidence supporting the existence of a particular structure is usually obtained by analyzing products generated by ribonucleases with known structure-specific activities (Knapp (1989)). The molecule under scrutiny is labelled at either its 5' or 3' terminal nucleotide with 32P, and then submitted to partial digestion by each of a battery of specific RNases (for the minimum number of enzymes, see Zaritsky & Forester (1991)), under conditions which are assumed not to interfere with the native RNA conformation. Lengths of the labelled fragments thus generated are determined by gel electrophoresis and autoradiography, with a ladder containing all alkaline hydrolysis products of the RNA as size markers. These lengths identify the positions of enzymatic cleavages, and hence the local secondary structure (Pieler et al. (1986); Gerhart et al. (1986); Ehresmann et al. (1987)). [If the actual sequence of the given RNA molecule is unknown, the same procedure can be exploited to determine it, but with base-specific single-stranded RNases operating in denaturing conditions.]

Deciphering the results of such experiments are however tedious and not always fruitful. Algorithms that would allow computers do this job are desirable but unavailable. Here, we attempt to present one (Section 3), which is based on the third, neural network approach (1.1)

8.4.2 THE 'REVERSE' (SIMULATION) APPROACH

A parameterizable simulation program (GEL) has recently been developed (Zaritsky & Forester (1991)), to aid in the analysis of the partial RNase digestion data by displaying ideal autoradiograms. The 5'-terminal 75 bases of murine J-chain mRNA with a high potential for forming secondary structures with presumed biological activities (Zaritsky et al. (1988); Ben-Alon (1989)), served as an example to demonstrate the applicability of GEL. The program simulates the pattern of migration on an electrophoretic gel of a set of denatured RNA fragments obtained by partial cleavage of the native structure by each of the selected enzymes. However, GEL can only predict experimental results if some structure is assumed, a structure that is obtained by modelling as discussed above (Section 1). If one of GEL's predictions is discordant with the corresponding experimental results (e.g., Knapp (1989)), the tested structure should be abandoned and another one may be introduced and tested; the procedure can be repeated until a particular predicted structure is confirmed.

GEL has two major disadvantages: (a) It can only confirm the existence of such a structure, not predict one itself; (b) The procedure to confirm one particular folding may be exceedingly time-consuming, because it depends on much luck in the order of choosing among the many possible alternative structures, predicted by theoretical considerations.

8.4.3 THE DIRECT, 'FORWARD' APPROACH

Instead of using the computer as a simulator for the autoradiogram from a hypothetical structure (Section 5) obtained by other procedures (1) as does GEL (Zaritsky & Forester (1991)), one can better exploit computer's potential in a more sophisticated way by having it analyze the data obtained from scanning an experimental autoradiogram and deduce the extant structure, rather than predict the autoradiogram.

8.4.3.1 SCANNING PROCEDURES

Processing of the autoradiogram is done in two phases. In the first, the autoradiogram is scanned and each lane is compared against the ladder lane. This results within a set of points, each with its single strand characteristics (and double strand, in native conditions only). This phase uses standard line-detection techniques to identify the RNA fragments and filter out noise lines (Rosenfeld & Kak, 1982). Similar techniques are used for DNA autoradiograms (Gray et al. (1984); Russell et al. (1984); Elder et al. (1986); West (1988)). The identification can be done by either processing each of the lanes separately or by analyzing them as one lane. The above process is repeated for several specific RNases so that as much information as possible is gathered. This results in a set of points along with the following attributes: a. Its distance from the beginning of the sequence (the base's number); b. The enzyme which 'discovered' the point and its 'coupling' characteristics.

The final outcome of this stage is represented by a base sequence of the relevant RNA molecule (first attribute), with the paired bases (second attribute) marked specifically (see e.g. the M format of Fig. 2 in Zaritsky & Forester (1991)).

8.4.3.2 DERIVING THE STRUCTURE

The aim of this stage is to transform the M format to a planar representation of the structure (e.g., the H (Hairpin) format of Fig. 2 in Zaritsky & Forester (1991); and see here Figs. 2a, 2d and 4c). This problem is analogous to the one solved for small RNAs by Takefuji et al., 1990b, and can therefore be incorporated within a neural network structure (Section 1.1) and handled by the MIS algorithm (2) or the maximum planar subgraph algorithm (3). The procedure of marking bases reduces the number of pairings and thus the number of bases involved in the structure, hence reducing the number of required processing elements in both algorithms. Consequently, it decreases the computation time. Information on the marked bases also provides an accurate prediction of the structure by reducing the number of valid alternatives: The chance to obtain the global optimum solution among many local minima is enhanced due to the smaller number of possible base pairings.

Ideally, and as a first approximation, all the marked bases in the molecule under study are paired, and the only paired bases are those marked; in other words, there is a unique correspondence between a mark and a pairing. [As a corollary, there should be an even number of marked bases in the resultant M format. Cases in which this condition is not fulfilled will be discussed briefly below (Section 5).]

8.5 ADDITIONAL PROBLEMS

One can expect several kinds of problems in this approach, because the experimental procedures are subject to various diversions from the ideal picture drawn (Section 1). The following are examples for which solutions can (a) or cannot (b) be easily incorporated: (a) Secondary recognition sites for enzyme cleavages; (b) Interactions of certain stretches of the studied RNA with any regulatory molecule or structure in the living cell (be it DNA, RNA, protein, low molecular weight molecules or a combination of these).

8.5.1 SECONDARY RECOGNITION SITES

This artifact results in weak bands on the autoradiogram, which can be treated rigorously, as for the following example: RNase T1 usually cleaves downstream (at the 3'-end of) any unpaired G and does not cleave paired Gs. However, lower cleavage efficiencies are ocasionally observed downstream a paired G when it has a 3' neighboring U which is not paired. The resultant weaker band can either be ignored by the image analysis procedure (Section 4.3.1), or it can be labeled as a second-type mark on the M format output (Zaritsky & Forester (1991); Section 4.2) and recognized by a simple addition to the algorithm (Sections 2 & 3; to be published elsewhere).

8.5.2 REGULATORY CELLULAR MOLECULES

Many in vivo regulatory signals involve interactions between macromolecules, low molecular weight molecules or structures. A pertinent example is the temporary binding of a ribosome to an mRNA, which is necessary for translation of the latter to a protein. The nature of this signal is not fully understood yet, and much effort is being expended to decipher this interaction, which is crucial for any living cell (e.g., Kozak (1986)). Any stretch of RNA involved in such binding is protected from cleavage(s) by the specific RNase(s) thus eliminating the relevant band(s) from the autoradiograms, bands which usually appear in the in vitro reaction mixtures (not commonly containing ribosomes). Such an interaction will be overlooked because these bands will show up in the absence of ribosomes. The only way to avoid this difficulty is by performing simulated 'real life' experiments (e.g., adding ribosomes to the mixture).

8.6 REFERENCES

Ben-Alon, D., 1989, Secondary structures in mRNA with biological activities. M.Sc. Thesis, Ben-Gurion University of the Negev.

Cimino, G.D., Gamper, H.B., Issacs, S.T. and Hearst, J.E., 1985, Psoralens as photoactive probes of nucleic acid structure and function: organic chemistry, photochemistry, and biochemistry. Annu. Rev. Biochem. 54, 1151-1193.

Dahlberg, J.E. and Abelson, J.N., 1989, RNA Processing (Part A: General Methods), Methods Enzymol. 180.

Dahlberg, J.E. and Abelson, J.N., 1990, RNA Processing (Part B: Specific Methods), Methods Enzymol. 181.

Diener T.O., 1987, The Viroid (Plenum Press, New York).

Ehresmann, C., Bandin, F., Mongel, M., Roinby, P., Ebel, J.P. and Ehresmann, B., 1987, Probing the structure of RNAs in solution. Nucl. Acids Res. 15, 9109-9128.

Elder, J.K., Green, D.K. and Southern, E.M., 1986, Automatic reading of DNA sequencing gel autoradiographs using a large format digital scanner. Nucl. Acids Res. 14, 417-424.

Fresco J.R., Alberts B.M. and Doty P., 1960, Some molecular details of the secondary structure of ribonucleic acid. Nature 188, 98-101.

Garrett, R.A., Douthwaite, S. and Noller, H.F., 1981, Structure and role of 5S RNA-protein complexes in protein synthesis. Trends Biochem. Sci. 6, 137-139.

Gerhart, E., Wagner, H. and Nordstrom, K., 1986, Structural analysis of an RNA molecule involved in replication control of plasmid R1. Nucl. Acids Res. 14, 2523-2538.

Goringer, H.U. and Wagner, R., 1988, 5S RNA structure and function. Methods Enzymol. 164, 721-747.

Gray, A.J., Beecher, D.E. and Olson, M.V., 1984, Computer based image analysis of one dimensional electrophoretic gels used for the separation of DNA restriction fragments. Nucl. Acids Res. 12, 473-491.

Gultyaev, A.P., 1991, The computer simulation of RNA folding involving pseudoknot formation. Nucl. Acids Res. 19, 2489-2494.

Holbrook, S.R., Sussman, J.L., Warrant, R.W. and Kim., S.-H., 1978, Crystal structures of yeast Phenyl-alanine Transfer RNA. J. Mol. Biol. 123, 631-660.

Holley L.H. and Karplus M., 1989, Protein secondary structure prediction with a neural network. Proc. Natl. Acad. Sci. USA 86, 152-156.

Jacobson, A.B., Kumar, H. and Zuker, M., 1985, Effect of spermidine on the conformation of Bacteriophage MS2 RNA. J. Mol. Biol. 181, 517-531.

Jaeger, J.A., Turner, D.H. and Zuker, M., 1989, Improved predictions of secondary structures for RNA. Proc. Natl. Acad. Sci. USA 86, 7706-7710.

Karplus, M. and Petsko, G.A., 1990, Molecular dynamics simulations in biology. Science 247, 631-639.

Knapp, G., 1989, Enzymatic approaches to probing of RNA secondary and tertiary structure. Methods Enzymol. 180, 192-212.

Kozak, M., 1986, Influence of mRNA secondary structure on initiation by eucaryotic ribosomes. Proc. Natl. Acad. Sci. USA 83, 2850-2854.

Le S-Y. and Zuker, M., 1991, Predicting common foldings of homologous RNAs. J. Biomolec. Struct. Dynam. 8, 1027-1044.

Major, F., Turcotte, M., Gautheret, D., Lapalme, G., Fillion, E. and Cedergren, R., 1991, The combination of symbolic and numerical computation for three-dimensional modeling of RNA. Science 253, 1255-1260.

Mankin, A.S., Skripkin, E.A., Ckichkova, N.U., Kopylov, A.M. and Bogdanov, A.A., 1981, An enzymatic approach for localization of oligodeoxynucleotides binding sites on RNA. FEBS Lett. 131, 253-256.

Martinez, H.M., 1988, An RNA secondary structure workbench. Nucleic Acids Res. 16, 1789-1798.

McCulloch W.S. and Pitts W.H., 1943, A logical calculus of ideas immanent in nervous activity. Bull. Math. Biophys. 5, 115-133.

Mei, H.-Y., Kaaret, T.W. and Bruice, T.C., 1989, A computational approach to the mechanism of self-cleavage of hammerhead RNA. Proc. Natl. Acad. Sci. USA 86, 9727-9731.

Noller, H.F., 1984, Structure of ribosomal RNA. Annu. Rev. Biochem. 53, 119-162.

Nussinov R., Pieczenik G., Griggs J.R. and Kleitman D.J., 1978, Algorithm for loop matching. SIAM J. Appl. Math. 35, 68-82.

Pieler, T., Guddat, U., Oei, S.L. and Erdmann, V.A., 1986, Analysis of the RNA structural elements involved in the binding of the transcription factor IIIA from Xenopus laevis. Nucl. Acids Res. 14, 6313-6326.

Pipas J.M. and McMahon J.E., 1975, Method for predicting RNA secondary structure. Proc. Natl. Acad. Sci. U.S.A. 72, 2017-2021.

Puglisi, J.D., Wyatt, J.R. and Tinoco, I. Jr., 1991, RNA pseudoknots. Acc. Chem. Res. 24, 152-158.

Qian N. and Sejnowski T., 1988, Predicting the secondary structure of globular proteins using neural network models. J. Mol. Biol. 202, 865-884.

Reid, B.R., 1981, NMR studies on RNA structures and dynamics. Annu. Rev. Biochem. 50, 969-996.

Rosenfeld, A. and Kak, A.C., 1982, Digital Picture Processing. Academic Press, New York, N.Y.

Russell, P.J., Crandall, R.E. and Feinbaum, R., 1984, Nucl. Acids Res. 12, 493-498.

Steeg, E.W., 1989, Neural network algorithms for RNA secondary structure prediction. Master's thesis, University of Toronto Computer Science Dept.

Stern S., Weiser B. and Noller H.F., 1988, Model for the three-dimensional folding of 16S ribosomal RNA. J. Mol. Biol. 204, 447-481.

Takefuji Y. and Lee K.C., 1989, A near-optimum parallel planarization algorithm. Science 245, 1221-1223.

Takefuji, Y., Chen, L.L., Lee, K.C. and Huffman, J., 1990a, Parallel algorithms for finding a near-maximum independent set of a circle graph IEEE Trans. Neural Networks 1, 263-267.

Takefuji, Y., Lin, C.W., and Lee, K.C., 1990b, A parallel algorithm for estimating the secondary structure in ribonucleic acids. Biol. Cybern. 63, 337-340.

Tinoco I., Uhlenbeck O.C. and Levine, M D., 1971, Estimation of secondary structure in ribonucleic acids. Nature 230, 362-367.

Turner, D.H., Sugimoto, N., Jaeger, J.A., Longfellow, C.E. Freier, S.M. and Kierzek, R., 1987, Improved parameters for prediction of RNA structure. Cold Spring Harhor Symp. Quant. Biol. 52, 123-133.

West, J., 1988, Automated sequence reading and analysis. Nucl. Acids Res. 16, 1847-1856.

Williams, A.L., Jr. and Tinoco, I., Jr., 1986, A dynamic programming algorithm for finding alternative RNA secondary structures. Nucl. Acids Res. 14, 299-315.

Zaritsky, A., Gollop, R. and Cann, G.M., 1988, Tertiary structure of the mRNA coding for J-chain might be involved in differentiation of mature B-lymphocytes to (IgM)5-secretory cells. Speculat. Sci. Technol. 11, 205-213.

Zaritsky, A. and Forester, E., 1991, A simulation program to display specific digestion products of predicted RNA foldings. Comput. Appl. BIOSciences 7, 57-60.

Zuker, M., 1989a, Computer prediction of RNA structure. Methods Enzymol. 180, 262-288.

Zuker M., 1989b, On finding all suboptimal foldings of an RNA molecule. Science 244, 48-52.

Zuker, M. and Stiegler, P., 1981, Optimal folding of large RNA sequences using thermodynamics and auxiliary information. Nucl. Acids Res. 9, 133-148.

Chapter 9

MISSIONARIES AND CANNIBALS PROBLEMS

AKIRA HANADA
YOSHIAKI ARAKI
YOSHIYASU TAKEFUJI
Keio University
Japan

KAZUHIRO TSUCHIYA
Fuji Electric, Japan

9.1 INTRODUCTION

In this paper, a novel parallel computing approach is presented where two applications including the missionaries-and-cannibals problem and the tower of Hanoi puzzle are used for demonstrating the effectiveness of the proposed approach. In the missionaries-and-cannibals problem, three missionaries and three cannibals have to cross the river to the opposite side where they must use a boat which holds at most two people. At no time, on neither bank, can the cannibals outnumber the missionaries, for then the missionary would be in danger. How do they manage to cross the river? This problem was introduced long time ago as recreational mathematic problem and has been studied by Simon and Newell (1) in 1961 using the cognitive method and by Bellman and Dreyfus (2) in 1962 using the dynamic programming method. The similar problem or "monkey problem" was introduced by McCarthy (3) in 1963. In general this problem with N missionaries, N cannibals, and the boat capacity B is called the M&C problem. The conventional tree search sequential method (4) requires $O(B^2 \bullet N^2)$ time. The space σ^* approach (5) (6) requires $O(B^2 \bullet N)$ time. The proposed parallel algorithm using $\frac{n^2}{2}$ processing elements requires four and only four steps, not depending on the problem size where n is given by $n=3N+1$. Note that the boat capacity B does not influence the number of required processors in our algorithm.

Fig. 1 Missionaries and cannibals problem.

Missionaries and Cannibals problems are often introduced in artificial intelligence. The most common approach uses the sequential tree-search method as shown in Fig. 2 for finding the successful path. In Fig. 2 the state of (3,3) indicates that the initial state of the problem has 3 missionaries and 3 cannibals on one side of the river. The path (state transition) from the state of (3,3) to (1,3) means that two cannibals cross the river. If one path-searching fails, the conventional method uses a backtracking scheme to find the different path. The computational complexity is determined by the complexity of the searched tree including the number of possible states. Although the tree-search method is able to provide all the possible paths, it cannot be used for solving large size problems since it takes a prohitively long time. The proposed parallel algorithm is not only able to find a solution as quickly as possible but also to find all possible paths.

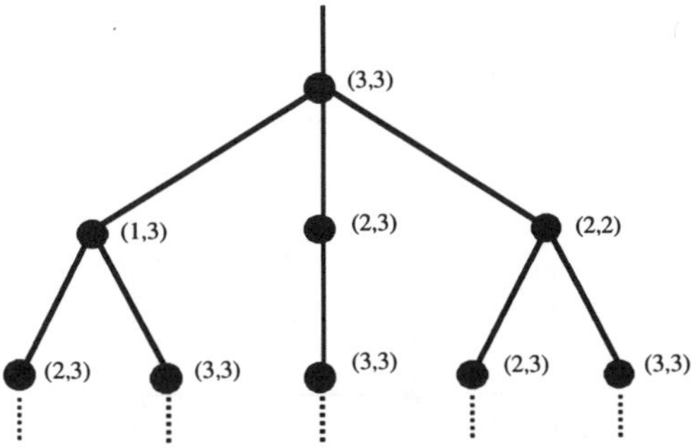

Fig. 2 Tree search for finding the successful paths.

The conventional parallel algorithms for the M&C problem are all based on transformation of the sequential method to parallel one where they did not take advantage of natural parallelism from the problem. The proposed parallel algorithm using $\frac{n^2}{2}$ processing elements requires four and only four steps where n is given by n=3N+1. Although the proposed algorithm is a parallel one, it was tested on several sequential machines where the system behaviors were simulated. The M&C problem with N=30, and B=4 was solved by the IBM-compatible-DOS 386-20MHz machine without coprocessor in 17 seconds, and the problem with N=50, and B=4 requires 79 seconds.

Saul Amarel (5) presented formulation of the state diagram called "space σ^* approach" for the M&C problem. Let c and m represent the number of cannibals and the number of missionaries on the first bank respectively. The ordered pair (c,m) denotes the state of the system with c cannibals and m missionaries on the first bank. The goal is to find a successful path from the ordered pair (3,3) to (0,0) where the following constraints must be always satisfied; 1) the number of cannibals should not be larger than that of missionaries on the boat and two banks, 2) one or more than one missionaries or cannibals must use a boat to cross the river. The state diagram of N=3 problem is illustrated in Fig. 3. Each white circle corresponds to a non-permissible state and black one for permissible state. In this problem the total of ten permissible states exist among 16 possible states. In general the total number of permissible states is given by (3N+1).

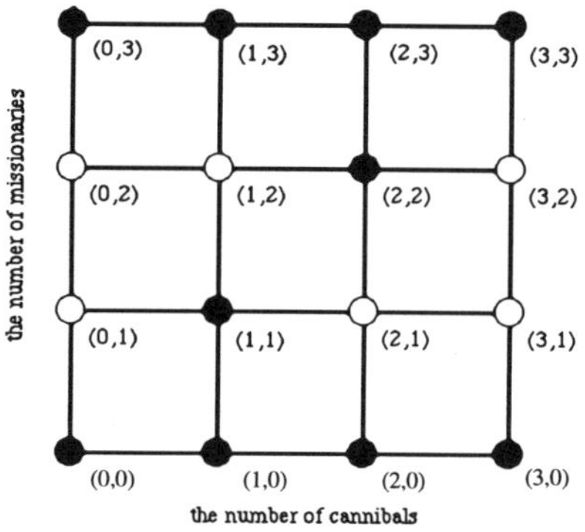

Fig. 3 The state diagram of N=3 problem by Amarel.

If the system is at the state (c_i,m_i) and the boat is on the first bank, then the permissible ordered pair (c_j,m_j) must satisfy $c_j \leq c_i$ (cannibal constraint),$m_j \leq m_i$ (missionary constraint) and $(c_i+m_i)-(c_j+m_j) \leq 2$ (boat capacity constraint). The state of the ordered pair (c_i,m_i) depicts that the state of the ith circle has c_i cannibals and m_i missionaries. If the system is at the state (c_i,m_i) and the boat is on the other bank, then the ordered pair (c_j,m_j) is called the permissible ordered pair if $c_j \geq c_i$, $m_j \geq m_i$, and $(c_j+m_j)-(c_i+m_i) \leq 2$ are satisfied. For example, the state transition (3,3)$\Rightarrow$(1,3), (3,3)$\Rightarrow$(2,3), or (3,3)$\Rightarrow$(2,2) satisfies the given constraints. In the Amarel's method, permissible paths are searched from the state diagram based on the constraints in all steps.

9.2 PARALLEL ALGORITHM

In our algorithm, the M&C problem is solved by $\frac{n^2}{2}$ processing elements where n is given by n=3N+1. Instead of using the ordered pair (c_i,m_i) in Fig. 3, our algorithm uses the new numbering scheme which plays a key role in parallel computing. We have two numbering-circle methods. One economical method is to number black circles from right to left and up to down which can save the required memory space considerably in simulation on the sequential machine. Using the method, the number of processing elements is $\frac{n^2}{2}$ where n is given by n=3N+1. The ith black circle of the ordered pair is numbered by the following equation: if m_i=N then $i=N+1-c_i$, if $m_i=0$ then $i=3N+1-c_i$, if $c_i=m_i$ and $m_i \neq N$ and $m_i \neq 0$ then $i=2N+1-c_i$. The other method is to number the black and the white circles which is suitable for parallel computing but requires the large memory space in sequential-machine simulation where the required memory space is given by $(N+1)^4$. The ith circle of the ordered pair (c_i,m_i) is numbered by the following equation: $i=(N+1)(N-m_i+1)-c_i$. The state of the ijth single processing element represents a path between the ith circle and the jth circle. $V_{ij}=1$ means that a path between the ith circle and the jth circle is permissible and $V_{ij}=0$ for non-permissible.

The proposed parallel algorithm is composed of four steps. In the first step the processing elements are all set to zeros. In the second step all possible paths are connected between the ith circle and the jth circle for i,j=1,..., 3N+1. If the path from the ith circle to the jth circle is permissible, then the ijth processing element will generate nonzero output ($V_{ij}=1$) else $V_{ij}=0$. The permissible constraints from the circle of (c_i,m_i) to the circle of (c_j,m_j) are given by $c_j \leq c_i, m_j < m_i$ and $(c_i+m_i)-(c_j+m_j) \leq 2$, and (c_i,m_i) and (c_j,m_j) must be black-colored. In the third step, the feasible paths will remain and the unacceptable paths will be removed to find a successful path or to find all the successful paths. In the last step, if the generated path is successful, there will be no change in the state of the system, else the state of the system will be changed which informs us unsolvability of the given problem.

Step-1: All outputs of the processing elements are set to zero: $V_{ij}=0$ for $i,j=1,...,3N+1$.

Step-2: The ijth processing element for $i,j=1,...,3N+1$ and $i>j$ is updated by:

$$V_{ij}=f(|m_i-m_j|+|c_i-c_j|,B)\,d((m_i-m_j)(c_i-c_j))g_i g_j \tag{1}$$

where $f(L,R)=\begin{cases}0 & \text{if } L>R\\ 1 & \text{otherwise}\end{cases}$ and $d(x)=\begin{cases}1 & \text{if } x\geq 0\\ 0 & \text{otherwise}\end{cases}$

and $g_i=\begin{cases}1 & \text{if } c_i=m_i \text{ or } m_i=0 \text{ or } m_i=N\\ 0 & \text{otherwise}\end{cases}$

The function of the boat capacity constraint $f(|m_i-m_j|+|c_i-c_j|,B)$ is 1 if the state transition between (c_j,m_j) and (c_i,m_i) is acceptable and 0 otherwise. The function of $d((m_i-m_j)(c_i-c_j))$ is 1 if the state on the boat among the transition between (c_j,m_j) and (c_i,m_i) is permissible and 0 otherwise. d(x) function is used for detecting the state of (missionary-increase and cannibal-decrease) or that of (missionary-decrease and cannibal-increase). g_i is 1 if the ordered pair (c_i,m_i) is permissible and 0 otherwise. $g_i g_i=1$ means that the state transition from the ith state to the jth state is permissible. These functions including f(x), d(y), and $g_i g_i$ are logically ANDed to generate the successful paths.

Step-3: Every path between two permissible ordered pairs must satisfy the reachability constraint. If a permissible ordered pair does not links from/to the other path it must be deleted, else it should remain. The state of every processing element describes the horizontal or the vertical path. The condition of the reachability constraint is given by:

$$V_{ij}=V_{ij}h\left(\sum_{\substack{k=1\\k\neq i}}^{n} V_{kj}\sum_{\substack{k=1\\k\neq j}}^{n} V_{ik}\right)$$

where $h(x)=\begin{cases}0 & \text{if } x=0\\ 1 & \text{otherwise}\end{cases}$

Step-4: Step-3 is repeated by once and only once.

$$V_{ij}=V_{ij}h\left(\sum_{\substack{k=1\\k\neq i}}^{n} V_{kj}\sum_{\substack{k=1\\k\neq j}}^{n} V_{ik}\right)$$

If one or any permissive ordered pairs in **Step-3** are removed from **Step-4**, then it indicates that no successful path exists in the problem.

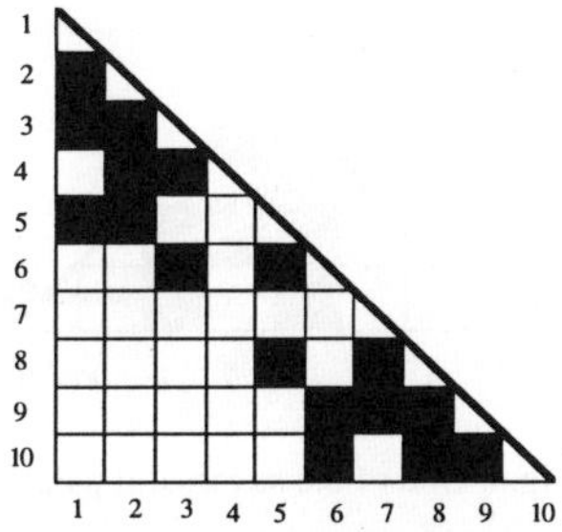

Fig. 4 The state of processing elements after Step-2.

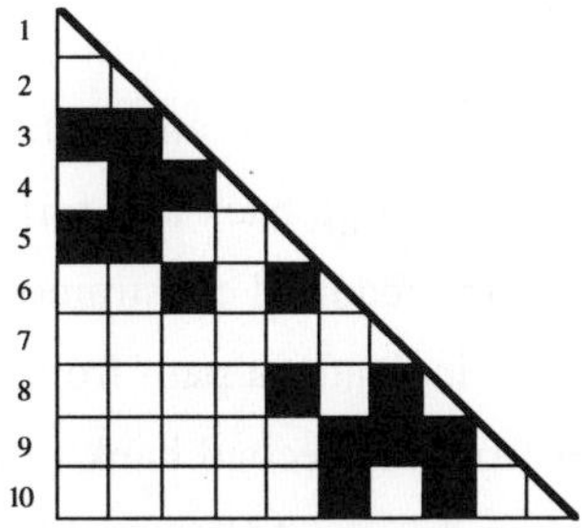

Fig. 5 The processing elements after Step-3.

Fig. 4 and Fig. 5 show the state of processing elements after Step-2 and Step-3 respectively. The black square indicates that $V_{ij}=1$, or the permissible path between the ith and the jth ordered pair exists. The white square means the non-permissible path in Fig. 4 and Fig. 5. As shown in Fig. 5 two states of (2,1) and (16,15) in Fig. 4 are removed after Step-3.

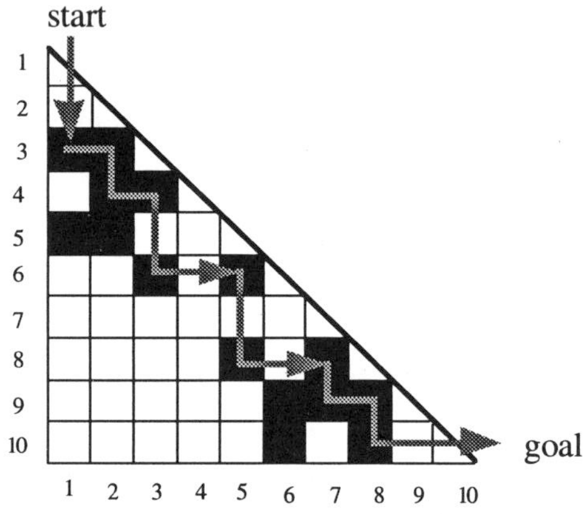

Fig. 6 One of the successful paths.

Fig. 5 includes all possible paths from the initial state to the final state. In order to quickly pick one successful solution from the all possible solutions in Fig. 5, the following is a simple procedure: Move a rook from the initial state #1 to the final state #10 where the rook must move row and column through black squares alternatively. It means that a boat must go back and forth between the first bank and the second bank under satisfying the required constraints. Every rook move of the ith state number on the black square indicates a path from the first bank to the second bank, and the jth state number from the second bank to the first bank. For example, Fig.6 shows one of the solutions for the M&C problem with N=3 and B=2. The successful path from circle number #1 to circle number #10 is given by; $1 \Rightarrow 3 \Rightarrow 2 \Rightarrow 4 \Rightarrow 3 \Rightarrow 6 \Rightarrow 5 \Rightarrow 8 \Rightarrow 7 \Rightarrow 9 \Rightarrow 8 \Rightarrow 10$. The successful path numbers can be easily translated into the state diagram, the path is $(3,3) \Rightarrow (1,3) \Rightarrow (2,3) \Rightarrow (0,3) \Rightarrow (1,3) \Rightarrow (1,1) \Rightarrow (2,2) \Rightarrow (2,0) \Rightarrow (3,0) \Rightarrow (1,0) \Rightarrow (2,0) \Rightarrow (0,0)$.

In general the translation from the circle numbers to the state diagrams is given by;

1) When the system is economical numbering method (using $(3N+1)^2/2$ processing elements):

if $1 \le i \le N+1$

mi=N, and ci=N+1-i

if $N+2 \le i \le 2N$

mi=ci=2N+1-i

if $2N+1 \le i \le 3N+1$

mi=0, and ci=3N+1-i

2) When the system is large numbering method (using $(N+1)^4$ processing elements):

m_i=N-truncate[(i-1)/(N+1)], and c_i=modulo[N+1- modulo (i,N+1), N+1]

where "truncate" is a function to truncate any fraction. For example, truncate(5/2)=2 and modulo(5,2)=1. Fig. 7, Fig. 8 and Fig. 9 indicate the path from the initial state to the final state. Note that (100,100) means 100 cannibals and 100 missionaries on the first bank. The path is given by from left to right and from up to down.

(4,4) (3,3) (3,4) (2,2) (3,3) (3,0) (4,0) (1,0) (2,0) (0,0)

Fig. 7 One permissible path for N=4, B=3 problem.

(5,5) (4,4) (4,5) (3,3) (3,5) (0,5) (2,5) (2,2) (3,3) (3,0) (4,0) (1,0) (2,0) (0,0)

Fig. 8 One permissible path for N=5,B=3 problem.

(100,100) (98,100) (99,100) (94,100) (95,100) (95,95) (96,96) (94,94) (95,95) (93,93) (94,94) (92,92) (93,93) (91,91) (92,92) (90,90) (91,91) (89,89) (90,90) (88,88) (89,89) (87,87) (88,88) (86,86) (87,87) (85,85) (86,86) (84,84) (85,85) (83,83) (84,84) (82,82) (83,83) (81,81) (82,82) (80,80) (81,81) (79,79) (80,80) (78,78) (79,79) (77,77) (78,78) (76,76) (77,77) (75,75) (76,76) (74,74) (75,75) (73,73) (74,74) (72,72) (73,73) (71,71) (72,72) (70,70) (71,71) (69,69) (70,70) (68,68) (69,69) (67,67) (68,68) (66,66) (67,67) (65,65) (66,66) (64,64) (65,65) (63,63) (64,64) (62,62) (63,63)

(61,61) (62,62) (60,60) (61,61) (59,59) (60,60) (58,58) (59,59) (57,57) (58,58) (56,56) (57,57) (55,55) (56,56) (54,54) (55,55) (53,53) (54,54) (52,52) (53,53) (51,51) (52,52) (50,50) (51,51) (49,49) (50,50) (48,48) (49,49) (47,47) (48,48) (46,46) (47,47) (45,45) (46,46) (44,44) (45,45) (43,43) (44,44) (42,42) (43,43) (41,41) (42,42) (40,40) (41,41) (39,39) (40,40) (38,38) (39,39) (37,37) (38,38) (36,36) (37,37) (35,35) (36,36) (34,34) (35,35) (33,33) (34,34) (32,32) (33,33) (31,31) (32,32) (30,30) (31,31) (29,29) (30,30) (28,28) (29,29) (27,27) (28,28) (26,26) (27,27) (25,25) (26,26) (24,24) (25,25) (23,23) (24,24) (22,22) (23,23) (21,21) (22,22) (20,20) (21,21) (19,19) (20,20) (18,18) (19,19) (17,17) (18,18) (16,16) (17,17) (15,15) (16,16) (14,14) (15,15) (13,13) (14,14) (12,12) (13,13) (11,11) (12,12) (10,10) (11,11) (9,9) (10,10) (8,8) (9,9) (7,7) (8,8) (6,6) (7,7) (5,5) (6,6) (4,4) (5,5) (5,0) (6,0) (1,0) (2,0) (0,0)

Fig. 9 One permissible path for N=100, B=5 problem.

9.3 THE TOWER OF THE HANOI PUZZLE

Using the same approach, we have solved the Tower of the Hanoi puzzle (7). Fig. 10 shows three pegs and six disks where every disk has a hole in the center and they are different in terms of size. First, they are piled up in order of their size in #1 peg. The goal is to move all disks to peg #3 where one disk can be only moved at a time and #2 peg can be used as temporary buffer. We must always satisfy that any larger disk should not be piled upon the smaller disk in three pegs. In general the problem with N disks is called Hanoi puzzle. The Hanoi puzzle is often introduced to explain recursive programming. There are many sequential methods for the Hanoi puzzle including the tree search, the A^{KT} search, the A* search, and so on (8,9). However no 'parallel' methods have been proposed as far as we know. The proposed parallel method requires n^2 processors where $n=3^N$. In order to find a successful path it requires two and only two steps.

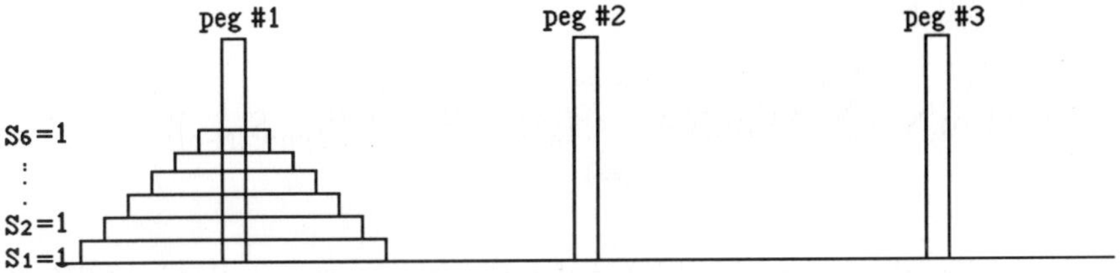

Fig. 10 Hanoi puzzle.

It is important that n x n processing elements are numbered as shown in Fig. 10 where they represent the location of the disks. In general the state of the system including the locations of the disks is given by $(S_{i1},S_{i2},...,S_{iN})$ where a single number i represents the state of $(S_{i1},S_{i2},...,S_{iN})$:

$$i = 1+\sum_{k=1}^{N} 3^{N-k}(S_{ik}-1)$$ where S_{ik} indicates the peg # of the kth disk in ith state.

For example, i=8 with three disks can be translated into the state of (1,3,2) where (1,3,2) informs us that the largest disk is on peg#1, the middle disk is on peg #3, and the smallest disk is on peg #2 . The all possible paths between permissible states can be generated by the following two-step parallel procedure:

Step-1: All outputs of the processing elements are set to zeros: $d_{ij}=0$ for $i,j=1,...,3^N$

Step-2: If the path from the ith state to the jth state is permissible, then $d_{ij}=1$, and else $d_{ij}=0$.

$$d_{ij}= f\left(\sum_{k=1}^{N} h\left(S_{ik}-S_{jk}\right)\right) * h\left(M(m,N)+\sum_{k=1}^{N-1} M(m,N-k), \prod_{p=0}^{k-1} \left(g\left(S_{im},S_{iN-p}\right)g\left(S_{jm},S_{jN-p}\right)\right)\right)$$

where f(x)=1 if x=1, 0 otherwise. h(x)=1 if x≠0, 0 otherwise. g(L,R)=1 if L≠R, and 0 otherwise. M(L,R)=1 if L=R, and 0 otherwise. if $h(S_{ik}-S_{jk})=1$ then m =k.

Fig. 11 shows the connectivity between permissible states for N=3 problem. The black squares indicate that $d_{ij}=1$. In other words the path from the jth state to the ith state is permissible if $d_{ij}=1$. One path from the initial state (1,1,1) to the final state (3,3,3) can be found in Fig. 11. To quickly find one successful path, the neural network parallel algorithm was used in this paper. The system uses an n x n neural network array for solving the Hanoi puzzle where the output of the ijth hysteresis McCulloch-Pitts neuron or V_{ij} is given by (10): $V_{ij}=1$ if $U_{ij}>$UTP (upper trip point), 0 if $U_{ij}<$LTP (lower trip point), and unchanged otherwise. Note that U_{ij} is the input of the ijth neuron.

The motion equation is given by:
if $d_{ij}=1$ then

$$\frac{dU_{ij}}{dt}= -A\left(\sum_{k=1}^{n-1} V_{ik}d_{ik}-1\right)-A\left(\sum_{k=2}^{n} V_{kj}d_{kj}-1\right) +Bh\left(\sum_{k=1}^{n-1} V_{ik}d_{ik}\right)+Bh\left(\sum_{k=2}^{n} V_{kj}d_{kj}\right)$$

if $d_{ij}=0$ then $\frac{dU_{ij}}{dt}=0$

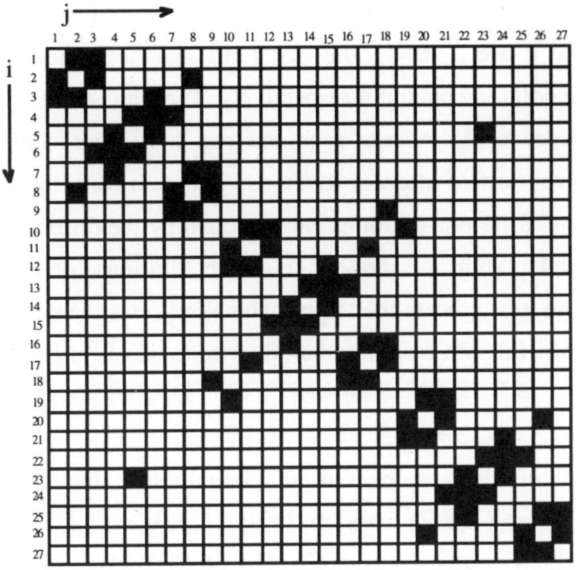

Fig. 11 The state of processing element after step-2.

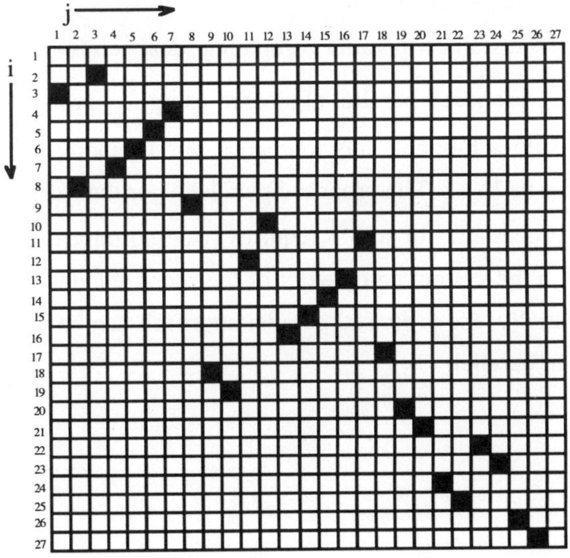

Fig. 12 One solution of Hanoi puzzle in Fig. 11.

where h(x) is called the hill-climbing term and h(x) is 1 if x=0, 0 otherwise. The coefficients A and B are A=B=1. The first term and the second term are the row constraint and the column constraint respectively. The first term forces one and only one neuron to fire per row in order to avoid the loop path. The second term forces one and only one neuron to fire in order to choose one path from the states. Using the proposed algorithm, different solutions can be found by the different initial uniform-random values of U_{ij}. Fig. 12 shows one of the solutions for N=3 Hanoi puzzle where the black squares indicate that $V_{ij}=1$ which indicates the path from jth state to ith state. The successful path from number #1 state to #27 is given by;

$$1 \Rightarrow 3 \Rightarrow 2 \Rightarrow 8 \Rightarrow 9 \Rightarrow 18 \Rightarrow 17 \Rightarrow 11 \Rightarrow 12 \Rightarrow 10 \Rightarrow 19 \Rightarrow 20 \Rightarrow 21 \Rightarrow 24 \Rightarrow 23 \Rightarrow 22 \Rightarrow 25 \Rightarrow 26 \Rightarrow 27.$$

The successful path numbers can be translated into the state diagram, the path is;

$$(1,1,1) \Rightarrow (1,1,3) \Rightarrow (1,1,2) \Rightarrow (1,3,2) \Rightarrow (1,3,3) \Rightarrow (2,3,3) \Rightarrow (2,3,2) \Rightarrow (2,1,2) \Rightarrow (2,1,3) \Rightarrow (2,1,1) \Rightarrow (3,1,1) \Rightarrow (3,1,2) \Rightarrow (3,1,3) \Rightarrow (3,2,3) \Rightarrow (3,2,2) \Rightarrow (3,2,1) \Rightarrow (3,3,1) \Rightarrow (3,3,2) \Rightarrow (3,3,3).$$

Fig. 13 shows the solution of the path.

This paper presents a new parallel computing approach for M&C problems and Hanoi puzzles. The proposed parallel algorithm using $\frac{n^2}{2}$ processing elements for M&C problems requires four and only four steps, not depending on the problem size where n is given by n=3N+1. The proposed parallel algorithm for the tower of Hanoi puzzle using n^2 processors for $n=3^N$ requires two and only two steps to find the solutions. Although the system complexity of the proposed Hanoi-puzzle algorithm increases exponentially, the proposed parallel approach can be used for solving the similar problems in parallel.

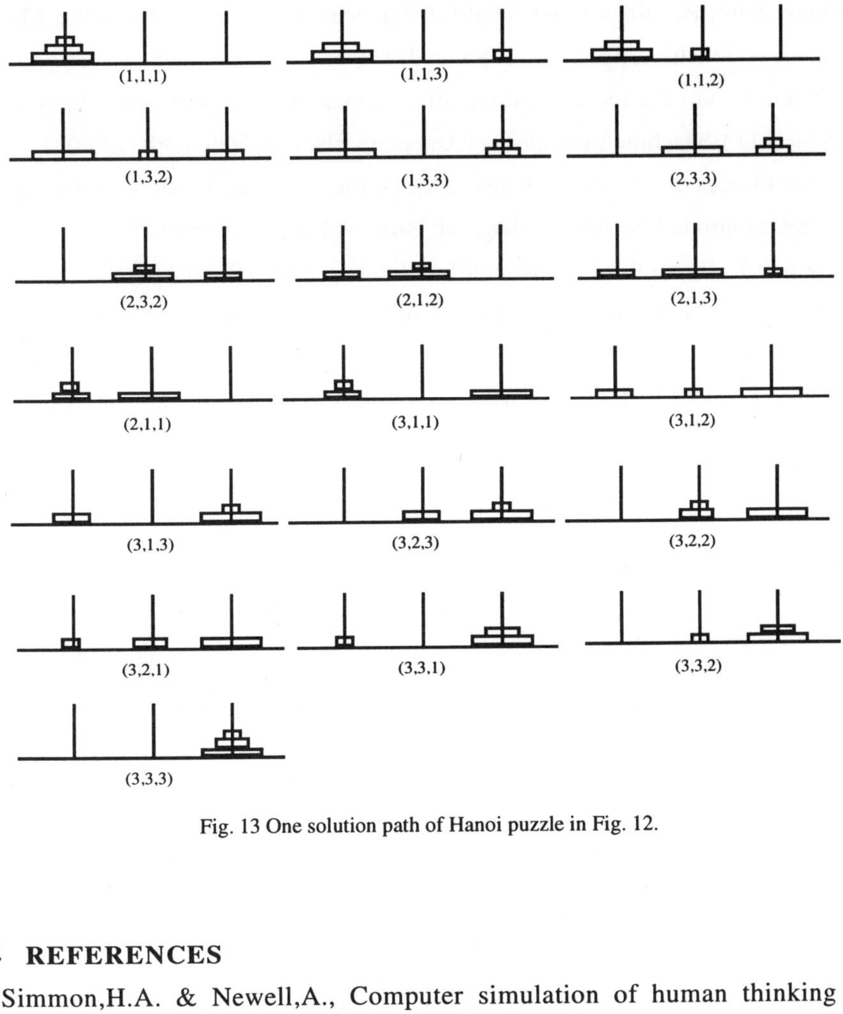

Fig. 13 One solution path of Hanoi puzzle in Fig. 12.

9.4 REFERENCES

1. Simmon,H.A. & Newell,A., Computer simulation of human thinking and problem solving. Datamation, June-july, 1961.

2. Bellmanm,R. & Dreyfus,S., Applied Dynamic Programming. Princeton: Princeton University Press, 1962.

3. McCarthy,J.,Situations, actions, and causal laws. Stanford Artificial Intelligent Project, Memo NO. 2, 1963.

4. Elaine Rich, Artificial intelligence, McGraw-Hill, 1991.

5. Amarel Saul, Readings in artificial intelligence : a collection of articles, Morgan Kaufmann, Publishers, pp. 2-22, 1985, c1981.

6. Amarel, S., On the representations of problems pf reasoning about actions, in: D.Michie(ED.), Machine Intelligence (American Elsevier, New york, 1968).

7. Edouard Lucas, Recreations mathematiques, four volumes, Gauthier-Villars, Paris, 1891-1894; reprinted by Albert Blancard, Paris, vol. 3 pp. 55-59, 1960.

8.Nilsson,N.J., Principles of Artificial Intelligence, Tioga Publishing Co. 1980.

9. Barr & feigenbaum (eds.), Principles of Artificial Intelligence, William Kaufmann. inc. 1981.

10. Takefuji Y. & Lee K. C. An hysteresis binary neuron: a model suppressing the oscillatory behavior of neural dynamics. Biological Cybernetics, 64, pp. 353-356, 1991.

Chapter 10

FUNCTIONAL-LINK NETS

YOSHIYASU TAKEFUJI
YOH-HAN PAO
Department of Electrical Engineering and Applied Physics
Case Western Reserve University
Cleveland, OH 44106, USA

10.1 INTRODUCTION

In 1985 Hopfield and Tank introduced an artificial neural network for obtaining solutions to complex combinatorial optimization problems. This new approach has attracted many new investigators to the field [1]. Subsequently Wilson and Pawley [2], and Paielli [3] cited difficulties attendant to the use of the Hopfield-Tank neural network for optimization problems and these findings have served to discourage support for neural network research for optimization problems. However Takefuji and his collaborators have reexamined a large number of well-formed optimization problems of academic and practical interest and have described how appropriately evolved versions of the original approach can indeed yield solutions superior to benchmark solutions [4].

Pao and his collaborators have advocated the practice of neurocomputing with enhanced representations [5] [6] for mapping, self-organization, and associative recall. To fix ideas we refer to that practice as functional-link net computing and acknowledge that the previous high-order nets of Giles and his collaborators are instances of that approach [7-9].

In our former paper [10] the functional-link net approach to ill-formed problems was briefly introduced. In this paper the mechanism is detailed and it is demonstrated that the functional-link net is very useful for solving ill-formed optimization problems where the analytical form of the Liapunov function is not known or not available. The bipartite subgraph problem which is NP-complete [11] was used to test the solution quality of the functional-link net..

10.2 FUNCTIONAL-LINK NET

As shown in Fig. 1, in physical terms, activation of nodes {k} by input pattern $\underline{X}$ offers the possibility that processes $f_0(\underline{X})$, $f_1(\underline{X})$,..., $f_n(\underline{X})$ may also be activated. The set of processes $f_j(\underline{X})$ are certainly likely to be nonlinear but need not be learned. Given input $\underline{X}$, an enhanced representation $\{f_j(\underline{X})\}$ is generated by the neural system and the mapping to be learned is that from $\{f_j(\underline{X})\}$ to the output $\{O_k\}$.

We emphasize that no intrinsically new information is introduced into the process. We merely volunteer the facts that if the input pattern vector is $\underline{X}$, then the value of $f_j(\underline{X})$ is such and such, and we generate in this manner an enhanced pattern representation of the input.

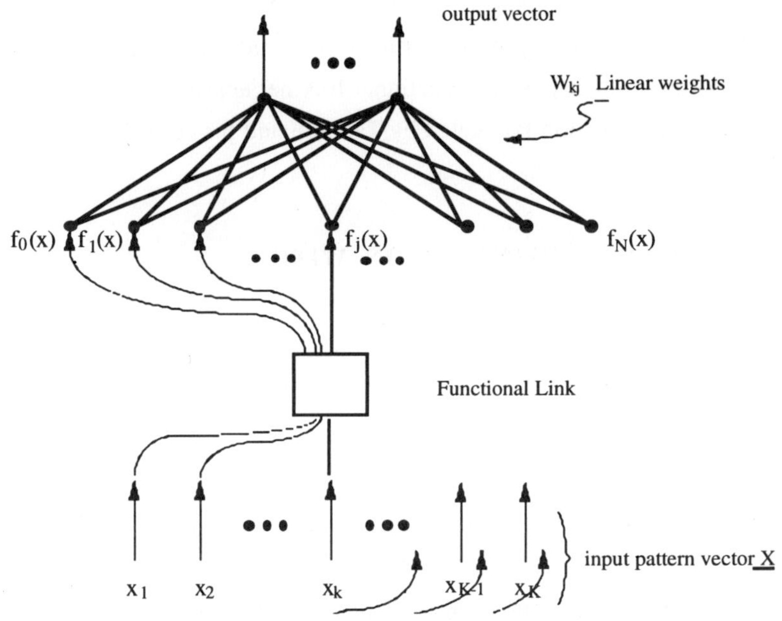

Fig. 1 Schematic of Functional Link Net.

10.3 OPTIMIZATION WITH THE FUNCTIONAL-LINK NET

The Hopfield and Tank neurocomputing approach to the computation of decisions in combinatorial optimization problems has stimulated great interest and has also received certain criticisms. The Hopfield-net approach depends on the availability of an approximate Liapunov function. For our purpose we might consider these to be "well-formed" problems. In other cases, the analytical form of the Liapunov function might not be known or not available and it might not be easy or advisable for the user to try to formulate an ad hoc Liapunov function. The motivation of using the functional-link net is to understand and learn the approximate Liapunov function for ill-formed problems. The functional-link net is very useful for learning a network representation of a Liapunov function and then the learned net is used to search for a configuration (solution) which corresponds to a local minimum in that function.

In this paper the bipartite subgraph problem which is NP-complete was used to demonstrate the functional-link net approach for ill-formed optimization problems. In order to show the solution quality of the functional-link net approach, the result of the best known Liapunov function fabricated by Takefuji and his collaborators [12] will be compared.

BIPARTITE SUBGRAPH OPTIMIZATION WITH FUNCTIONAL-LINK NET

The bipartite subgraph problem which belongs to NP-complete problems is defined by:

Instance: Graph G=(V, E), positive $K \leq |E|$.

Question: Is there a subset $E' \subseteq E$ with $|E'| \geq K$ such that G'=(V,E') is bipartite?

Consider a graph G with 5-vertex and 7-edge where G=(V, E) is given by V={ V_1, V_2, V_3, V_4, V_5} and E={ (1, 2) (1, 3) (1, 4) (2, 4) (2, 5) (3, 4) (3, 5)} as shown in Fig. 2 .

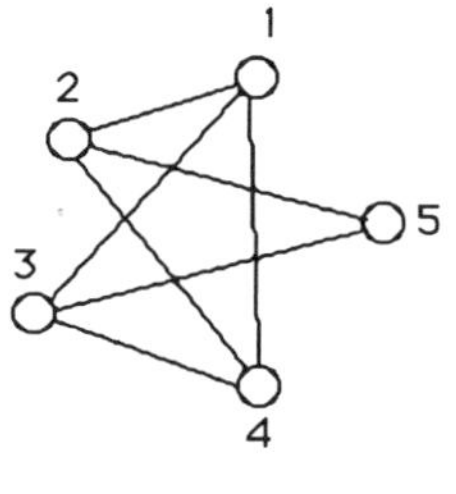

Fig. 2 Five-vertex problem.

One of the bipartite subgraphs G'=(V,E') is shown in Fig. 3 where an edge (1, 4) is removed from the original set of E.

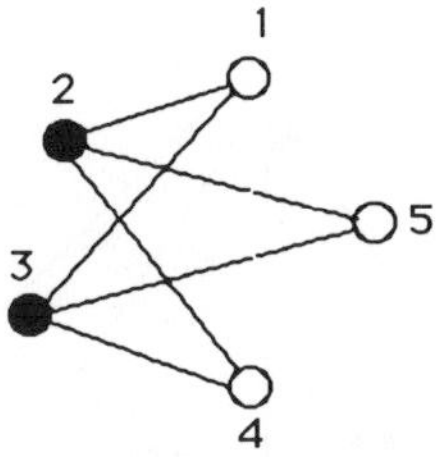

Fig. 3 One of bipartite subgraphs in Fig. 2.

Fig. 3 shows that one set of $\{V_2\ V_3\}$ with black colored and another set of $\{V_1\ V_4\ V_5\}$ with white colored without the edge (1, 4) construct the bipartite subgraph. The goal of the bipartite subgraph problem is to divide a graph into two subsets so as to minimize the number of removed edges where edges in the same subset are only removed from the given graph. Note that edges bridging between two subsets are not removed.

The state of the *i*th vertex whether it belongs to one subset or to another subset is represented by V_i=1 or V_i=0. The input/output function of the *i*th McCulloch-Pitts neuron is given by:

$$V_i = \begin{cases} 1 & \text{if } x_i > \text{Threshold} \\ 0 & \text{otherwise} \end{cases} \tag{1}$$

where x_i is the input of the *i*th neuron.

The optimization procedure with the functional-link net is composed of two subtasks. The first subtask is to learn an approximate Liapunov function $f(\underline{X})=net_p$ as shown in Fig. 4 for the bipartite subgraph problem and the second is to search for a local minimum of the Liapunov function. Let the unknown Liapunov function be the second order form:

$$net_p = \sum_j W_j g_j(x_1) + \sum_j W_j h_j(x_2) + \sum_j W_j s_j(x_1 \cdot x_2) \tag{2}$$

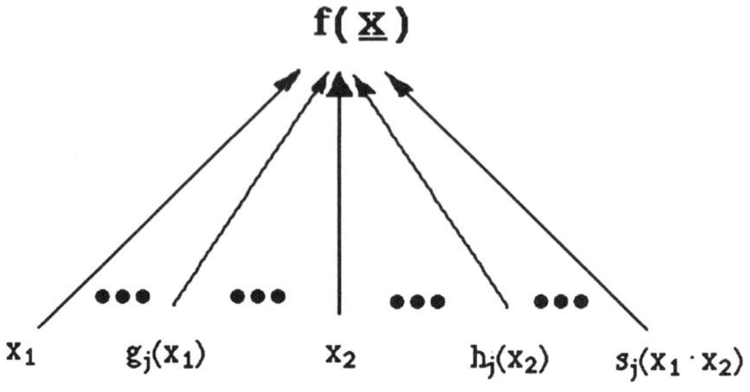

Fig. 4 Unknown Second Order Liapunov Function.

If we assume that $g_j(x_1)=x_1$, $h_j(x_2)=x_2$, and $s_j(x_1 \cdot x_2)=x_1 \cdot x_2$, the total input net_p fed to the functional link net will be simply given by:

$$net_P = \sum_i x_i \cdot g_i + \sum_i \sum_j x_i x_j W_{i,j} \tag{3}$$

The following two equations are used to learn the approximate Liapunov function to fix the weights g_i and W_{ij} of the functional-link net:

$$\frac{\partial W_{i,j}}{\partial t} = - \frac{\partial E_P}{\partial W_{i,j}} \tag{4}$$

and

$$\frac{\partial g_i}{\partial t} = - \frac{\partial E_P}{\partial g_i} \tag{5}$$

respectively where E_p is given by:

$$E_p = (net_P - penalty_p)^2. \tag{6}$$

The goal of the problem is to minimize the computational energy E for a given set of input patterns {P}. The function E is given by:

$$E = \sum_{P} E_P = \sum_{P} (net_P - penalty_p)^2 \tag{7}$$

The $penalty_p$ function returns the number of removed edges from the original graph to construct a bipartite subgraph.

Theorem 1: For weights g_i and W_{ij} learned in accordance with Eqs. (4) and (5), the computational energy E decreases with respect to time t.

Proof: Consider the derivatives of the computational energy with respect to time t.

$$\frac{dE}{dt} = \sum_{p} \frac{dE_p}{dt} = \sum_{p} \left(\sum_{i} \sum_{j} \frac{\partial W_{i,j}}{\partial t} \frac{\partial E_P}{\partial W_{i,j}} + \sum_{i} \frac{\partial g_i}{\partial t} \frac{\partial E_P}{\partial g_i} \right)$$

From Eq. (4) and (5)

$$\frac{dE}{dt} = - \sum_{p} \left(\sum_{i} \sum_{j} \left(\frac{\partial E_P}{\partial W_{i,j}} \right)^2 + \sum_{i} \left(\frac{\partial E_P}{\partial g_i} \right)^2 \right) \leq 0 \quad \text{Q.E.D.}$$

After all the weights are learned and fixed, we can search for a state of the system correspondence to a local minimum of the learned function net_p. The following motion equation for the *i*th neuron describes how the system evolves in accordance with gradient descent to obtain the system configuration (i.e. the graph configuration) corresponding to a local minimum of the learned Lyapunov function:

$$\frac{\partial x_i}{\partial t} = - \frac{\partial net_P}{\partial x_i} = - g_i - \sum_{j} W_{i,j} \cdot x_j \tag{8}$$

That configuration would also correspond to a graph with as many edges embedded as possible or as few edges removed as possible for a bipartite subgraph. However that optimal solution in general would be a local optimum.

Lemma: Eq. (8) states that the system converges to the local minimum of the Liapunov function net_p in Eq. (3).

Proof: Consider the derivatives of the function net_p with respect to the time t.

$$\frac{\partial net_P}{\partial t} = \frac{\partial x_i}{\partial t} \cdot \frac{\partial net_P}{\partial x_i} = -\left(\frac{\partial x_i}{\partial t}\right)^2 \leq 0 \qquad \text{Q.E.D.}$$

Fig. 5a shows the 10-vertex-27-edge graph problem. Fig. 5b and 5c show the final solution and the evolution of the system respectively. 15 input patterns which were randomly chosen from 1024 possible patterns were fed to the functional-link net to learn the approximate Liapunov function. Using the learned net and Eq. (8) the state of the system converged to a local minimum. In the final solution nine edges are removed from the original graph with 27 edges to construct a bipartite subgraph. In Fig. 5c the vertical axis (energy) represents the value of the function of net_p.

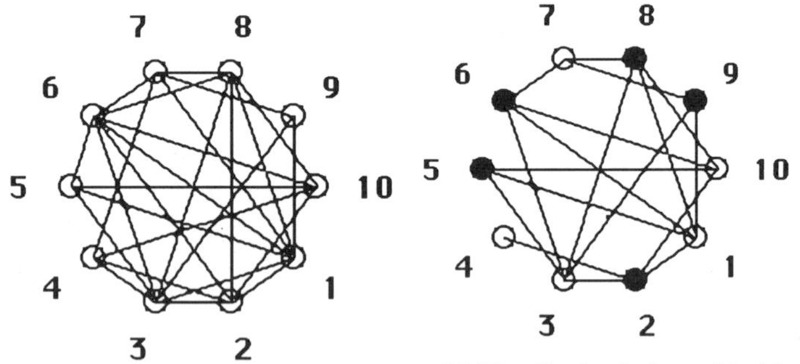

(a) The original graph

(b) The final solution with 18 edges embedded

Fig. 5(a)–(b) 10-vertex graph and its bipartite subgraph with the functional-link net.

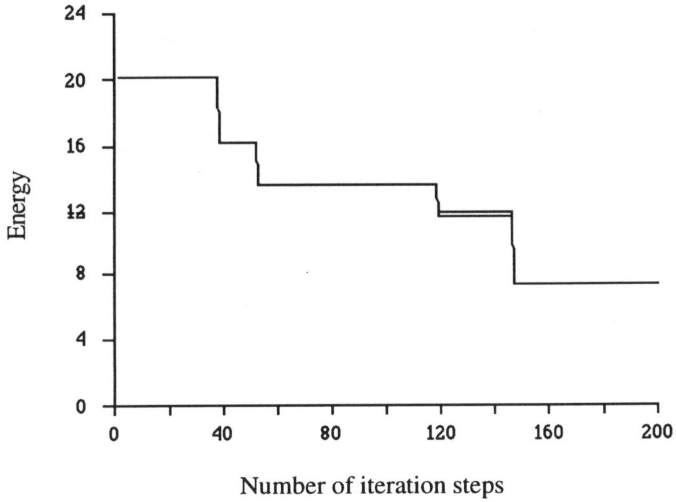

(c) The evolution of the optimization

Fig. 5(c) Simulation result with the functional-link net.

Based on our previous result [12] using the best known energy function E with the maximum neurons in Eq. (9), 20 edges were maximally embedded to construct a bipartite subgraph as shown in Fig. 6.

$$E = \frac{C}{2}\sum_{x=1}^{N}\sum_{y\neq x}^{N}\sum_{i=1}^{2} \text{Connection}(x,y)\, V_{x,i}\, V_{y,i} \tag{9}$$

where $V_{x,i} = 1$ if $U_{x,i} = \max\{U_{x,1}, U_{x,2}\}$, 0 otherwise. If $U_{x,1} = U_{x,2}$, the maximum neural network sets $V_{x,1} = 1$ and $V_{x,2} = 0$. $U_{x,i}$ and $V_{x,i}$ are the input and the output of the maximum neuron representing the *x*th vertex in the *i*th cluster assignment. Note that Connection(x,y)=1 if there exists an edge between vertex *x* and vertex *y*, 0 otherwise.

This simulation result in Fig. 5 and the previous result in Fig. 6 show that the functional-link net learned a good approximate Liapunov function from a few examples (15 patterns out of 1024 possible patterns) and it can be used for solving ill-formed problems.

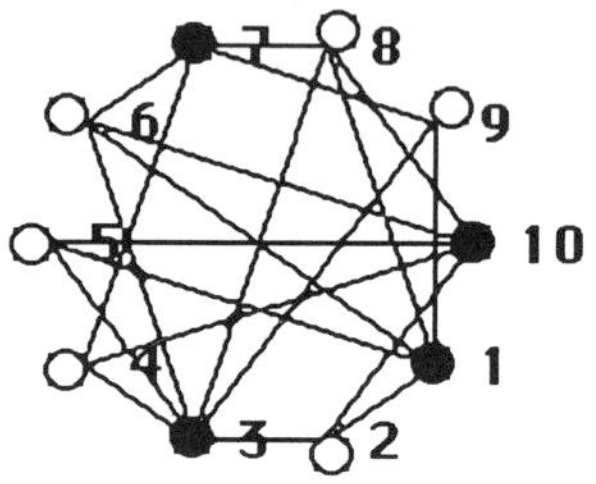

Fig. 6 The best solution by the best known energy function.

10.4 CONCLUSION

Although the solution obtained with use of a learned Liapunov function with 18 embedded edges is inferior to our best solution obtained otherwise with 20 embedded edges, we forward this example as an illustration of how to deal with ill-formed problems where the analytical form of the Liapunov function is not available. We believe that this additional feature extends the scope of utility of the Hopfield-Tank neural network approach to the computation of solutions to combinatorial optimization problems.

10.5 REFERENCES

[1] J. J. Hopfield and D. W. Tank, "Neural computation of decisions in optimization problems," Biological Cybernetics, 1985, 52, pp. 141-152.

[2] Wilson G. V. and Pawley G. S., (1988), On stability of the Travelling Salesman Problem Algorithm of Hopfield and Tank. Bio. Cybern., 58, 63-70.

[3] Paielli R. A., (1988), Simulation tests of the optimization method of Hopfield and Tank using neural networks, NASA Technical Memorandum 101047.

[4] Y. Takefuji, (1992) Neural Network Parallel Computing, Kluwer Academic Publishers.

[5] Y. H. Pao, "Adaptive pattern recognition and neural networks," Addison-Wesley, Reading, MA, 1989, Chapter 8, pp. 197-222.

[6] Y. H. Pao, "Functional-link nets: removing hidden layers," AI Expert, 4, Miller Freeman Pub., San Francisco, CA, 1989, pp. 60-68.

[7] H. H. Chen, Y. C. Lee, G. Z. Sun, H. Y. Lee, T. Maxwell, and C. L. Giles, "Higher order correlation model for associative memory," American Institute of Physics Conference Proc., No. 151:Neural networks for computing, Snowbird, UT, 1986, pp. 398-403.

[8] Y. C. Lee, G. Doolen, H. H. Chen, G. Z. Sun, T. Maxwell, H. Y. Lee, and C. L. Giles, "Machine learning using a higher order correlation network, Physics 22D, 1986, pp. 276-306, North-Holland, Amsterdam.

[9] C. L. Giles and T. Maxwell, "Learning, invariance and generalization in higher order neural networks, Applied Optics, 26, 1987, pp. 4972-4978.

[10] Y. H. Pao and Y. Takefuji, "Functional-Link Net Computing: theory, system architecture and functionalities," COMPUTER, May 1992.

[11] M. R. Garey and D. S. Johnson, Computers and Intractability : A Guide to the Theory of NP-Completeness, (W. H. Freeman and Co., San Francisco, 1979).

[12] K. C. Lee, N. Funabiki, and Y. Takefuji, "A parallel improvement algorithm for the bipartite subgraph problem," IEEE Trans. on Neural Networks, 3, 1, 1992.

Chapter 11

IDENTIFICATION AND CONTROL

STEPHEN M. PHILLIPS
CHRISTOPH MÜLLER-DOTT
Department of Electrical Engineering
Case Western Reserve University
Cleveland, Ohio 44106, USA

Neural network based strategies for real-time identification and feedback control of dynamic systems is presented. By coupling a real-time nonlinear system identification with a nonlinear control law an adaptive control paradigm for nonlinear systems is achieved. The system identification is based on a prediction error method which takes advantage of the general nonlinear approximation properties of neural networks. The accompanying issues of persistent excitation and identifier convergence are discussed in the context of successful identification for control design. For the nonlinear control a neural network implementation of a model based nonlinear one step ahead controller is used. This simple control strategy can be extended to other control laws and successful implementation is verified through a simulation example.

11.1 INTRODUCTION

The application of neural networks to control problems offers an alternative to traditional techniques. An architecture with separate networks for the identification and control of a nonlinear plant is developed. In addition, traditional nonlinear control methods are reviewed. The performance of the neural network technique is verified through simulations for the control of a nonlinear plant.

In this section we briefly review some frequently used traditional approaches to control nonlinear systems and then introduce the use of neural networks in control as an alternative. In Section Two we review several types of neural networks, including single-layer and multi-layer neural network structures, and we specifically point out some of the difficulties involved with their use in control. In Section Three we address the control problem in general and then show the configurations for the system identification and the self learning controller with the use of neural networks. The Fourth section describes and introduces the architecture for a specific simulation example. Section Five contains the results for the control system simulation example.

11.1.1 TRADITIONAL APPROACHES TO CONTROL NONLINEAR SYSTEMS

Continuous-time dynamic systems are typically modeled by ordinary differential equations. Frequently it is written as system of coupled first order differential equations, $\dot{\mathbf{x}} = \mathbf{F}(\mathbf{x}, \mathbf{u}, t)$, where the vectors **x** and **u** are the state and input to the system, and **F** is some nonlinear vector function. The goal of the control is to find the input, **u**, which provides the system with some desirable characteristics, for example stability and tracking performance. In the case of feedback control the input will be a function of some measurements of the system, for example **u=K(x)**.

A wide variety of control system design tools exist for designing linear controllers for linear systems. This is not the case for nonlinear systems or controllers. Although several techniques for control problems that include nonlinearities exist, they are not extensively used in practice due to their complexity. Another difficulty is that no general scheme is available to handle all nonlinear problems. Most methods use approximations, since exact solutions to nonlinear differential equations are seldom available. Among the most frequently used techniques are simulation, local linearization and gain scheduling, global feedback linearization, describing functions, equivalent gains and Lyapunov's second method.

Simulation typically begins by introducing a differential equation model for an actual or proposed system. The behavior of the modeled system can be revealed by using a numerical solution of the model for a digital computer simulation. Simulation does not give proof of stability, but provides confidence that under the given conditions the system performance is acceptable.

Local linearization is one of the most common techniques for nonlinear systems analysis and design, since it approximates the nonlinear system by a linear system for some operating region. Since there exists a large number of design and analysis tools for linear systems, the linearization approach simplifies the set of nonlinear differential equations and the approximate set of linear differential equations describing the system can be used. This local linearization technique becomes difficult for highly nonlinear systems since an unreasonable number of operating points may be required in order to accurately model the nonlinear system. For control this is especially important since the control design must be repeated for each linear model.

Feedback Linearization is a more global approach which requires a nonlinear controller which can remove the nonlinear behavior of the plant. This technique requires significant knowledge of the plant and essentially requires the construction of a nonlinear feedback controller to cancel the nonlinearities of the plant. The design of such a controller is nontrivial and the effects of plant mismodelling and disturbances are not completely understood [9].

There are other approximation techniques that are less frequently used but worthy of mention. The method of describing functions considers the response of the system to sinusoidal inputs [13]. The describing function consists of the gain and phase of the fundamental frequency of the system response and ignores all other components of the response. This allows an approximate stability and performance analysis of the nonlinear system [19]. Another technique is the method of equivalent gains. Here, a memoryless nonlinearity is replaced by a linear gain that varies between two extremes. Again, an approximate analysis of the system is tractable [5].

Lyapunov's second method [27] [18] is a tool to analyze the stability of dynamic systems on a theoretical basis. The method was developed to examine the stability of nonlinear differential equations. A very important feature of Lyapunov's second method is that it does not require the solutions of the nonlinear system equations. Although the method saves us the task of solving the system's differential equations, it does require the generation of a Lyapunov Function, which may be just as complicated and elusive. The idea behind Lyapunov's method is that a physical system can only store a finite amount of energy. If it can be shown that energy is always being dissipated, except at an equilibrium point, then the system must approach that equilibrium point. The generalized energy function for the system is called the Lyapunov Function.

Each of these conventional nonlinear control design techniques has drawbacks. It is because of these difficulties that we have proposed a neural network based alternative for nonlinear control.

11.1.2 USE OF NEURAL NETWORKS

Neural networks have the potential for very complicated behavior and their ability to learn is one of their main advantages over traditional nonlinear control techniques. The massive interconnections of the single processing units (neurons) in the multilayer networks provide the tool for neural network models [24]. Their significant fault tolerance and the capability for parallel processing were the initial impetus for interest in neural networks. Neural networks are currently used for pattern recognition [20], decision making [14] and fuzzy logic [12] as well as in control.

Most of the work done in the area of control has concentrated on either the control or the system identification part. Multilayer networks [23] or Hopfield nets [7] have been primarily used in this area. Shoureshi [26] and Atkeson [1] report successful implementation of system identification using associative memories such as the Hopfield net in a control system. Haesloop [6] uses a multilayer neural network for process identification of a dynamic system in his paper.

Wang [28], Patrikar [22] and Scott [25] used multilayer neural networks for their self-learning or self-organizing controllers. A very detailed approach for system identification and control with neural networks has been given by Narendra [16]. Narendra demonstrated that neural networks can be used effectively for the identification and control of nonlinear dynamic systems. Other publications for system identification and control include work by Nguyen [17] and Chen [3] paper. All of this work was done using multilayer neural network structures. Even with all of their potential advantages it should be noted that artificial neural networks should not be applied to all control problems; they should rather be understood as another tool for control design and analysis and used to expand the designer's choices for attacking demanding control problems.

11.2 ARTIFICIAL NEURAL NETWORKS

Artificial Neural Networks consist of a number of elemental processing units which are connected to one another. The *inputs* to the unit $x_1, \ldots\ldots, x_n$ are multiplied by the corresponding *weights* $w_1, \ldots\ldots, w_n$. The result then is summed and the *activation function F* finally transforms this weighted sum into the output *o*.

$$o = F(\Sigma_i x_i w_i) = F(\mathrm{x}^T \mathrm{w}) = F(net) \tag{1}$$

The weights represent the strength of a connection for the input x_i. Since large structures with many processing units are common in artificial neural networks, inputs need not only be external environment inputs but they can also represent outputs from preceding units.

The activation function F(net) can have a variety of different shapes. The most commonly used shapes include the familiar linear, ramp, step and sigmoidal functions. The sigmoid function is

$$F\ (\text{net}) = \frac{1}{1 + e^{-\frac{net + \theta}{\theta_o}}} \tag{2}$$

where the parameters θ *and* θ_o change the slope of the activation function and allow a shifting along the x-axis. One important characteristic is that the first derivative

$$F(\text{net}) = F(\text{net})\ (1 - F(\text{net}))/\theta_o \tag{3}$$

simplifies gradient descent learning algorithms and therefore the sigmoid activation function is most commonly used.

11.2.1 MULTILAYER NETWORKS

Multilayer networks, as introduced by Rumelhart, Hinton and Williams [23] consist of a number of processing units arranged in different layers. In general, a multilayer network is made up of one input and one output layer and at least one hidden layer. The outputs of the processing unit in one layer are transmitted to nodes in the next layer through the weights, which amplify or attenuate the outputs. Except for the input layer nodes, the input to each node of the net is the sum of the weighted outputs of the prior layer nodes. The shape of the activation function, the input and the bias of each node determine the final output of the processing unit.

11.2.2 SUPERVISED LEARNING

In supervised learning, pairs of associated input and output patterns are learned using the Generalized Delta Rule with backpropagation of error algorithm introduced by Rumelhart et al. [23]. The network tries to capture the functional characteristic from the input layer to the output layer for a number of different patterns.

During the learning phase of training a multilayer network, we present the input vector $\underline{I}_p$ (p denotes the different patterns) as the input to the net and we want the net to adjust its weights, thresholds and biases such that the output vector $\underline{O}_p$ reaches a desired target value

$\underline{T}_p$. We wish to accomplish the capture of this input/output relationship for all patterns, i.e. all input patterns should generate the corresponding output value with a single set of thresholds and weights.

In training the error of the network for a particular pattern is

$$E_p = \frac{1}{2} \sum_n (t_{pn} - o_{pn})^2 \tag{4}$$

where n denotes the length of the output vector.

The average system error over all patterns can be defined as [20]

$$E = \frac{1}{2\,P} \sum_p \sum_n (t_{pn} - o_{pn})^2 \tag{5}$$

It is the task of the generalized delta rule (GDR) [23] to adjust the weights, thresholds and biases such that the error E_p or the average error E is minimized. This is achieved by using a gradient descent algorithm in weight space. The error is calculated from top to bottom, that is the algorithm starts at the output layer and back propagates the error to the input layer in order to find the necessary changes in weight space to reduce the overall error. See [20] and [23] for details.

It is important to recognize that the GDR algorithm does not succeed in every case. Satisfying results may take thousands of iterations due to a slow convergence rate and the remaining error might still be large. An even more troublesome problem with the GDR is that the error functions in (Equation 4) or (Equation 5) which have to minimized can be very complicated, since the output of the multilayer network includes nonlinearities and it is determined by a large number of nodes and layers. The error function is not quadratic so the gradient descent algorithm in weight space will not necessarily reach the absolute minimum. As with any gradient algorithm a local but not global minimum may be reached. Significant also is the determination of the number of layers and the number of nodes per layer needed to achieve the best results and convergence. These difficulties have prompted the investigation of other networks.

11.2.3 THE FUNCTIONAL LINK NET

The Functional Link Net (FLN) is a higher order net proposed by Pao [21] in 1988 to improve the performance of the multilayer networks. In the GDR net, the processing unit is a simple one in which the input is given by $\Sigma w_n \; x_n$. Thus the connecting links form the input as a linear combination of the outputs of the nodes from the previous layer. In the FLN, nonlinear connection links are introduced; for example a multiplicative nonlinear connection combines the output of two or more processing units and multiplies them before entering the sum of the next node. Pao refers to the multiplicative connection as higher-order links. These nonlinear connection links are used as an extended input layer. The extended input layer is then directly connected to the output layer with linear links. Figures 1 and 2 show typical architectures for the functional link net.

In the GDR the network develops the internal representation of training sets and it creates a complex nonlinear mapping from the input to output layer, it therefore used hidden layers. In the FLN, user defined nonlinearities are created in the input space, so that hidden layers become unnecessary. Zwingelstein gives a mathematical justification for the use of the FLN in [29]. This justification relies on Hornik et. al. [8] where it is shown that the following approximation holds. For any continuous nonconstant function G mapping R to R, any continuous function $f(\underline{x})$ can be approximated as closely as desired by:

$$f(\underline{x}) \approx \sum_{j=1}^{q} \beta_j \left(\prod_{k=1}^{t_j} G\left(\sum_{i=1}^{r} w_{jki} x_i + b_{jk} \right) \right) \quad (6)$$

where q and t_j are integers and β_j, w_{jki} and b_{jk} are real constants.

Zwingelstein used Hornik's approach and demonstrated that the single layer FLN is capable of approximating any nonlinear function if the correct enhancements are present. Finding the correct enhancements on a trial and error basis is possible but very time-consuming. It is advisable to start with a large number of nonlinear enhancements and to prune the unnecessary terms after convergence to a single set of weights and thresholds is achieved. A more systematic approach can include the use of Taylor or Fourier series as terms for the enhancements [8].

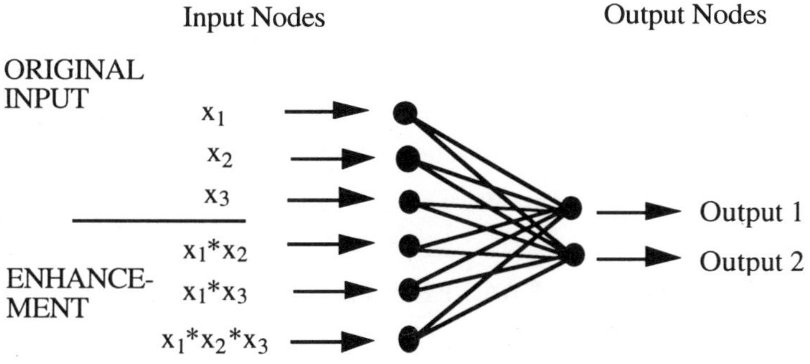

Fig. 1 Functional Link Net with Multiplicative Enhancements .

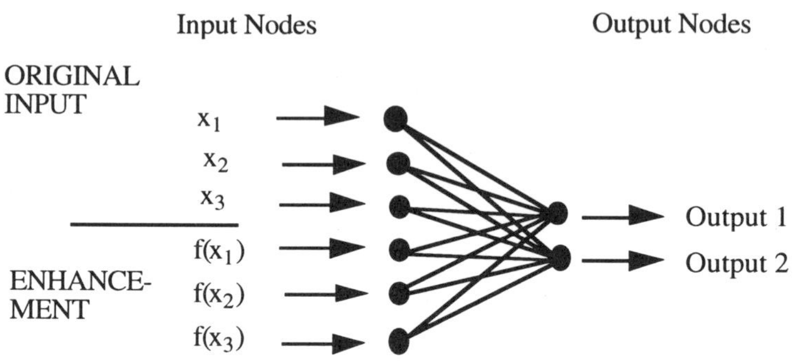

Fig. 2 Functional Link Net with Functional Enhancements.

11.2.4 SUPERVISED LEARNING WITH THE FUNCTIONAL LINK NET

The task of the supervised learning with the FLN is to capture the functional relationship and behavior of input/output pairs with the single layer net. The learning algorithm is similar to that of the GDR net. After the user has chosen the enhanced inputs, learning involves the change of the weights connecting the inputs and the enhanced inputs with the output layer nodes, until a minimum error is reached.

During the learning phase associated input/output pairs are presented to the net and the weights are adjusted to reduce the difference between the target and the output value. The delta rule [23] (gradient descent algorithm) used for this process determines the change in weights.

11.2.5 SPECIAL CHARACTERISTICS OF THE FLN

One of the major concerns with the multilayer networks is the local minima problem and the slow convergence speed. Since the FLN is a single processing layer net only, fewer calculations must be performed and no complex nonlinear structure has to be formed. Generally, this significantly reduces run time thus increases convergence speed. More importantly, the local minima problem, frequency a problematic feature of the GDR net, can be mitigated.

If for example, a linear activation function is chosen, then the output o_n of the FLN is

$$o_n = \left(\sum_i w_{ni} \; x_i \right) \tag{7}$$

which simply is the summation of the products of the inputs and the weights. Hence the error function

$$E = \frac{1}{2}(t_n - o_n)^2 \tag{8}$$

is a true quadratic function in weight space and the derivative with respect to the weights can be found from

$$\frac{\partial E}{\partial o_n} = -(t_n - o_n) = -\left(t_n - \sum_i w_{ni} \; x_i \right) \tag{9}$$

$$\frac{\partial E}{\partial w_{ni}} = \frac{\partial E}{\partial o_n} \frac{\partial o_n}{\partial w_{ni}} = -\left(t_n - \sum_i w_{ni} \; x_i \right) x_i \tag{10}$$

which guarantees that the absolute minimum for the error function in weight space will be found.

Using the nonlinear sigmoid activation in the processing unit can still generate local minima. Jones [11] conducted a number of experiments using linear and nonlinear activation functions and his reports indicate, that in general problems using the sigmoid activation function can be ruled out, especially if the saturation levels of the sigmoid activation are avoided. This is an important characteristic which allows an important advantage over the multilayer network structure.

11.3 THE CONTROL PROBLEM

In controller design the general idea is to find a controller that modifies a given plant's behavior to meet some objective. The overall goal is to find the plant input u which will be generated by a controller to cause the plant to behave in a desired way.

Our goal is to find controllers for plants whose dynamics are unknown or poorly known and possibly nonlinear. The task of constructing a controller for an unknown plant is divided into two parts.

- Modeling of the unknown plant, i.e., system identification
- Constructing a self-learning controller using an objective function defined by the designer. The two steps are described in detail in the sequel.

11.3.1 SYSTEM IDENTIFICATION

System identification is the process of constructing models and estimating unknown plant dynamics from measured data. A model which describes the system's behavior is useful in order to design a controller for the dynamic system. Once a good model for the true plant has been found, simulations and experiments are facilitated enormously.

For our purposes a number of assumptions will be made.

- The plant can include linear or nonlinear system dynamics
- All states that are needed to control the plant must be observable
- The plant is a time-invariant system

Since we do not require *a priori* knowledge of the plant for system identification, parametric methods are of no use since we cannot predict the order of the system or the form of any parametric model. For our approach we will use pairs of input and output measurements taken from the original plant to build the model. **u**(t) and **y**(t) represent vector signals corresponding to the plant inputs and measured states respectively. Initial inputs **u**(t) are presented to the system and the outputs **y**(t) are measured to receive the corresponding input/output pairs. Note that the states represent all of the past information, or memory of the plant. This is consistent with the usual linear control definition of state. It is the task of the model, in this case a neural network, to construct a representation of the plant using the given input/output pairs taken from the real plant. This is shown in Figure 3 where we desire the output estimates $\hat{\mathbf{y}}(t)$ to be close to the real outputs **y**(t).

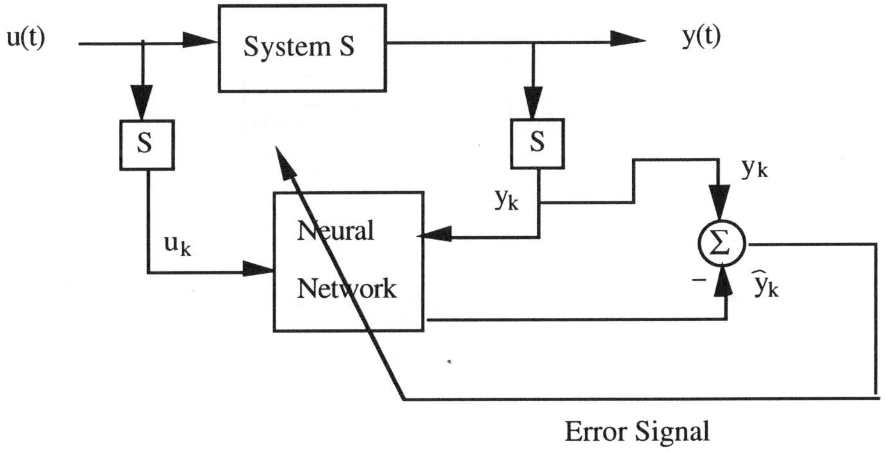

Fig. 3 Learning for System Identification.

The structure of the neural network for the system identification can either be a multilayer feed-forward network or a single layer network such as the functional link net. In the case of a nonlinear system S and the use of the functional link net, the correct nonlinear enhancements have to be found so that the mapping from input to output space, i.e. the learning of the functionality of the plant, can be achieved. The learning process for the model to match the system S is done by minimizing the error function.

$$J = (y(t) - \hat{y}(t))^2$$

for continuous time systems and respectively

$$J = (y_k - \hat{y}_k)^2$$

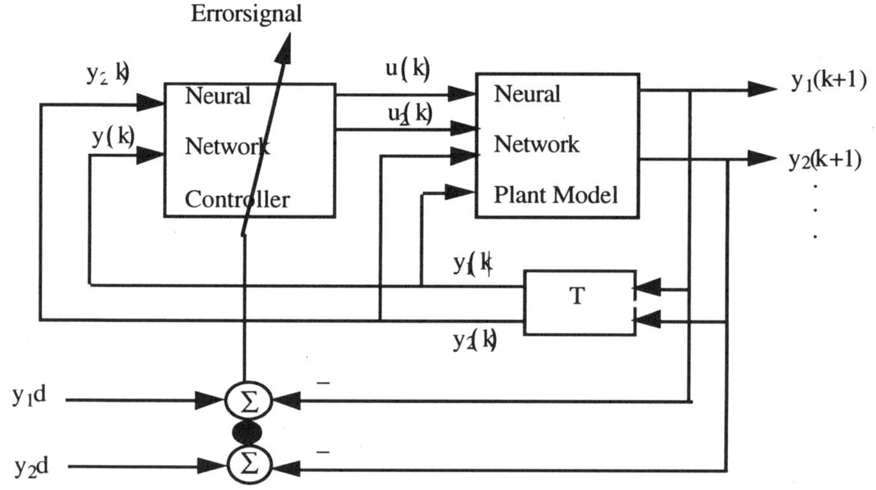

Fig. 4 Self-learning controller.

for sampled discrete time systems for every $\mathbf{u}(t)$ or any $\mathbf{u}_k$. As shown in Figure 4, $\mathbf{u}_k$ is an input vector derived from the input $\mathbf{u}(t)$ by sampling with a sampled period T.

The training for the model is straightforward. The sampled data pairs of $\mathbf{u}_k$ and $\mathbf{y}_k$ are the patterns presented to the neural network. The error signal is calculated and the weights and thresholds inside the neural network are adjusted to match $\hat{\mathbf{y}}(t)$ to $\underline{\mathbf{y}}$ for every input $\mathbf{u}$.

Training is terminated when the remaining system error is small enough and the neural network model can then be used as a model for the real plant.

11.3.2 SELF-LEARNING CONTROLLER

Given that the system identification part has been successful we use the neural network model of the true plant in the controller design. Using the model instead of the real plant gives us more freedom and makes simulation and the neural network design simpler.

The task of the controller is to generate a control signal that will make the plant behave in a desired manner. For this example we want the controller to drive the plant to a desired final state while tolerating disturbances and inaccurate data. Since the neural-net model will not be exact the controller must accommodate inaccuracies in the plant model.

As in the system identification, either a multilayer or a single layer neural network structure can be chosen for the controller. The number of inputs and outputs for the controller depends on the number of plant inputs and the number of measured states coming from the model. Note that this structure is very similar to the standard state feedback control design for linear plants except that now both the plant model and state feedback control are nonlinear (Figure 4).

The error signal is used to train the controller according to the control objective. The error function

$$J = \alpha_1 \left[y_{1d} - y_1(k+1)\right]^2 + \alpha_2 \left[y_{2d} - y_2(k+1)\right]^2 + \ldots \tag{11}$$

has been used in this work although other control criteria could be chosen. The weights and the thresholds in the neural network controller are then adjusted to minimize this error. The coefficients α weight each error depending on its importance.

The training process and training data are acquired in the following way. We start by presenting an initial state $\mathbf{y}_k$ to the trained neural network model of the plant. In addition the control $\mathbf{u}_k$ is also presented to the plant model. After the next sample we measure the plant model output at time $k + 1$, i.e. $\mathbf{y}_{k+1}$. Since the neural network controller is untrained, the control presented to the model will not have the correct value, so that $\mathbf{y}_k$ does not evolve into the desired state $\mathbf{y}_d$ at time $k + 1$. The error signal is used to train the neural network which will change its weights and thresholds to output the correct values for the controller state $\mathbf{u}_k$. The training takes place until $\mathbf{y}_k$ and $\mathbf{u}_k$ map into $y_{k+1} = y_d$. At this point the first complete pattern has been acquired. This is essentially a version of the one-step-ahead control strategy.

A new set of initial states is then presented to the plant model and the process is repeated, to find the corresponding control to achieve the desired state. The number of different initial inputs used to train the controller determines the accuracy and the range of the controller. The training can be done either after all patterns have been collected or at every time a new pattern is acquired. In the latter case we simply add the new pattern to the existing training patterns and reiterate the training process. The procedure is an automatic learning process and the controller design using this method only requires us to specify the objective function and to use the error signal to find the correct controller output. Therefore the name self learning controller.

11.4 Simulation Example

In order to show the capabilities of this approach we implement the neural structures for control purposes and show the performance in a simulation example. We choose an unstable nonlinear plant with good simulation characteristics. The unstable plant will require an effective controller for any reasonable performance.

11.4.1 The Inverted Pendulum

The inverted pendulum control problem is often used as an example of an inherently unstable dynamic system to demonstrate various control techniques. Barto et al. [2] showed that neuronlike elements with their learning capabilities are able to solve difficult learning control problems. Here we will use the inverted pendulum to verify our control strategy.

Figure 5 shows the pole cart system as an extension to the inverted pendulum, where

F = control force

m_C = mass of the cart

l = length of the pole to its center of mass

Θ = angle displacement from vertical axis

m_P = mass of the pole

A rigid pole is mounted on the top of a wheeled cart. The cart can freely move on a straight track. The pole is free to move and fall in one vertical plane only. The goal is to balance the pole at a certain angle by applying forces to the cart. The forces will move the cart to the left or right depending on the force values and friction is not modeled. Michie and Chambers [15] as well as Chen [4] also studied the pole cart system using controllers based on neural network structures. Their approaches differ significantly from ours. Michie and Chambers used bang-bang control dividing up the control space in a large number of local decision making elements to find the control output. Chen used the pole-balancing problem to demonstrate control based on system inverse models. In another approach Iwata [10] used fuzzy control to balance the pole and to make it follow a given trajectory.

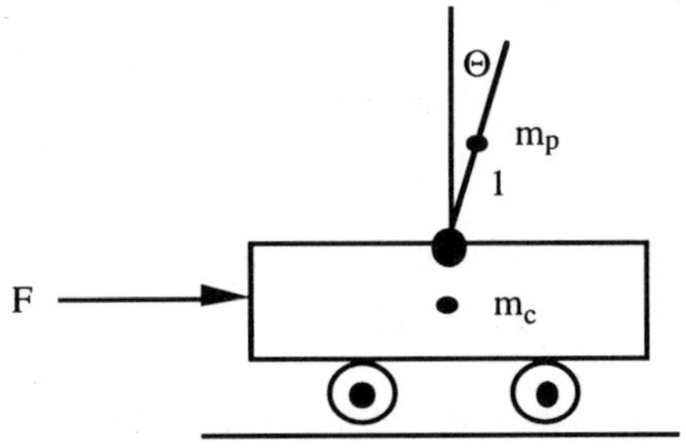

Fig. 5 The pole and cart example.

11.4.2 SIMULATION MODEL FOR THE INVERTED PENDULUM

The dynamic equation for the pole care system [2] with the assumption of zero-friction is

$$\ddot{\Theta} = \frac{g\sin\Theta + \cos\Theta\left[\dfrac{-F - m_p l\dot{\Theta}^2 \sin\Theta}{m_c + m_p}\right]}{l\left[\dfrac{4}{3} - \dfrac{m_p \cos^2\Theta}{m_c + m_p}\right]} \tag{12}$$

A fourth-order Runge-Kutta-Fehlberg integration with a time step of $0.1s$ was used to simulate Equation 12. The first and second derivatives of the angle Θ were calculated by the central difference method. The parameters in Equation 12 for the pole care system were selected to be

$$m_c = 1.0kg$$
$$m_p = 0.1kg$$
$$l = 0.5m$$
$$g = 9.81 \frac{m}{s^2}$$

11.4.3 SYSTEM IDENTIFICATION

Since the neural network plant model used for the system identification part is a fixed memoryless nonlinear function, the correct choice of states is the key to a good plant model. The system consists of two states, the angle Θ , the velocity $\dot{\Theta}$ as well as the input control force F. The output of the system are the angle Θ and the velocity $\dot{\Theta}$. The input/output data pairs for each initial condition are used to train the neural network model. The training is successful if the neural network estimates $\hat{\mathbf{y}}_{k+1}$ closely match the true values of $\mathbf{y}_{k+1}$.as in Figure 6.

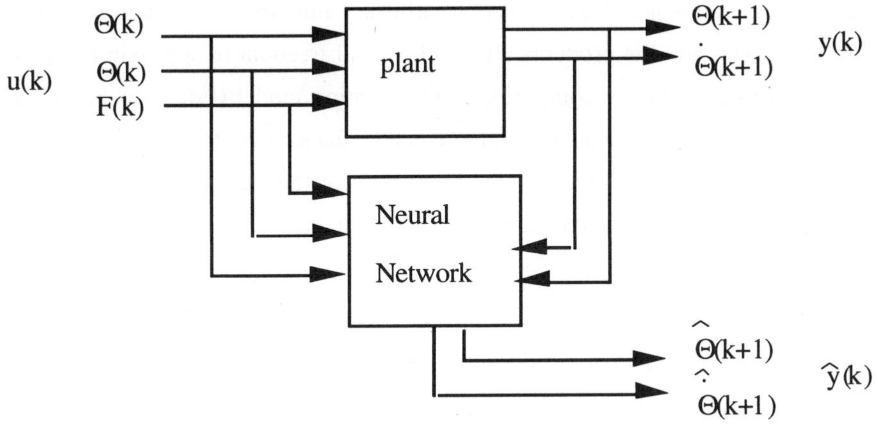

Fig. 6 System identification with the neural network model.

All neural network structures were designed using NNet210 (NNet210 is a product of AI Ware Inc.). For the system identification part a GDR-net and a single layer FLN design were implemented. The criterion for the system identification was to minimize the error between the true and estimated values of Θ and $\dot{\Theta}$, i.e.

$$min\ J = \left[\left(\Theta - \dot{\Theta} \right)^2 + \left(\dot{\Theta} - \hat{\dot{\Theta}} \right)^2 \right] \tag{13}$$

The pattern learning file which was generated during the simulation contains 287 patterns covering the desired learning space. Combinations of the pattern elements chosen for training are listed in the Table 1. The pattern file is the same for both structures.

The GDR-net has two hidden layers and 12 nodes in each hidden layer. Training for all patterns was completed after 1046 iterations (see Figure 7) and the remaining system error is 0.00005. The single layer functional link net has 37 enhancements; both functional and outer product enhancements are used to allow significant freedom in the mapping from input to output space. This choice of enhancements is fairly generic. The sinusoids are chosen since an arbitrary function can be approximated using Fourier series and the products allow additional freedom in approximating the model. Note in particular by comparing the simulation model (Equation 12) to the choice of enhancements (Appendix) that there is no enhancement that exactly matches the model. The success of this example shows that for this case the neural network approach is robust to the particular choice of enhancements. The predicted results for $\Theta\ (k+1)$ and $\dot{\Theta}(k+1)$ result in a system error of 0.00003. 231 iterations were needed to accomplish acceptable training results (see Figure 7). The 37 enhancements are listed in the appendix.

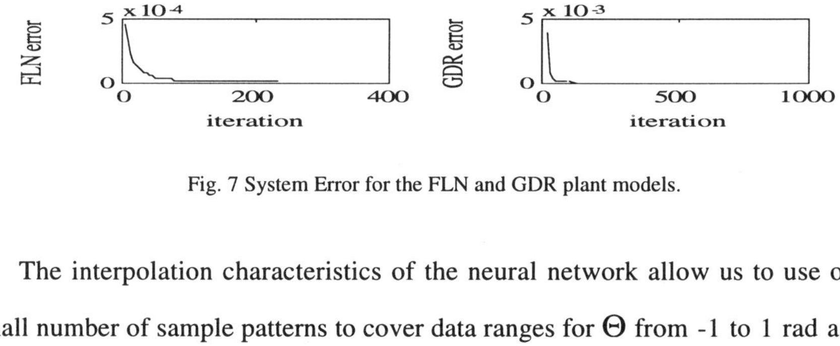

Fig. 7 System Error for the FLN and GDR plant models.

The interpolation characteristics of the neural network allow us to use only a small number of sample patterns to cover data ranges for Θ from -1 to 1 rad and $\dot{\Theta}$ from -5 to 5 rad/s and F from -50 to 50 N. After training, NNet allows us to consult the trained network, i.e. we can enter inputs and monitor the network's output. It should be mentioned that normalization techniques allow control over the desired output range of values. The normalization ranges for the pole cart system were selected to be

$$-80N \leq F \leq +80N$$

$$-5\frac{rad}{s} \leq \dot{\Theta} \leq +5\frac{rad}{s}$$

$$-1.5rad \leq \Theta \leq +1.5rad$$

11.4.4 THE CONTROLLER DESIGN

For the controller design a second neural network with the FLN structure was chosen. The controller has two inputs coming from the neural network model of the plant, Θ and $\dot{\Theta}$.

One control output, the control force, F, is needed. During the training phase different enhancements were tested and the final 11 enhancements chosen are also listed in the appendix.

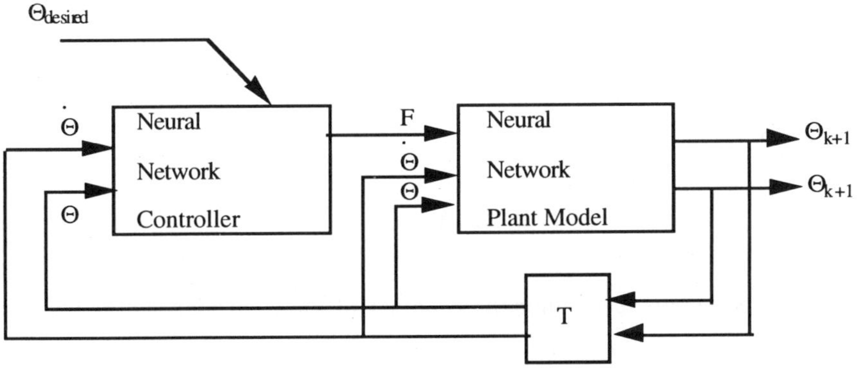

Fig. 8 Neural Network Controller Design.

Figure 8 shows the controller architecture. 29 patterns spread evenly over the range of data to control the inverted pendulum were used to train the controller. The elements chosen for the patterns are listed in Table 2. After 1572 iterations the remaining system error was 0.00001 (Figure 9). The objective in our example was to control the pole to the vertical (zero-angle) position and data was collected according to the objective function.

$$\Theta_{desired} = 0 \tag{14}$$

Every sample the states Θ and $\dot{\Theta}$ are presented to the controller which produces the corresponding control force F to attempt to achieve zero angle. This is repeated at each time step until Θ and $\dot{\Theta}$ approach zero. In order to find the values for the outputs of the controller and the plant model, each of the neural networks has to be consulted alternately.

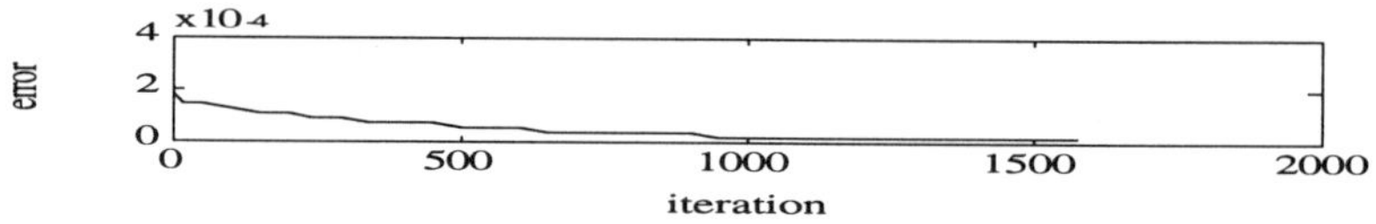

Fig. 9 System Error for the FLN.

11.5 RESULTS

11.5.1 CONTROL SYSTEM BEHAVIOR

For the first experiments to monitor the behavior of the controller in combination with the plant model, no disturbances or noise were added. Different initial states have been chosen and the system states were stored. The neural networks of the controller and the plant model are consulted with the present data and the outputs are presented in a feedback loop to the system.

Figure 10 displays the control force and position along the time samples for an initial condition of $\Theta = 0.3$ and $\dot{\Theta} = 0$. Θ, $\dot{\Theta}$ and the control force F all reach approximately zero and balance the pole back to the vertical axis. At the final samples only small control forces are required to compensate the inherently unstable dynamic system.

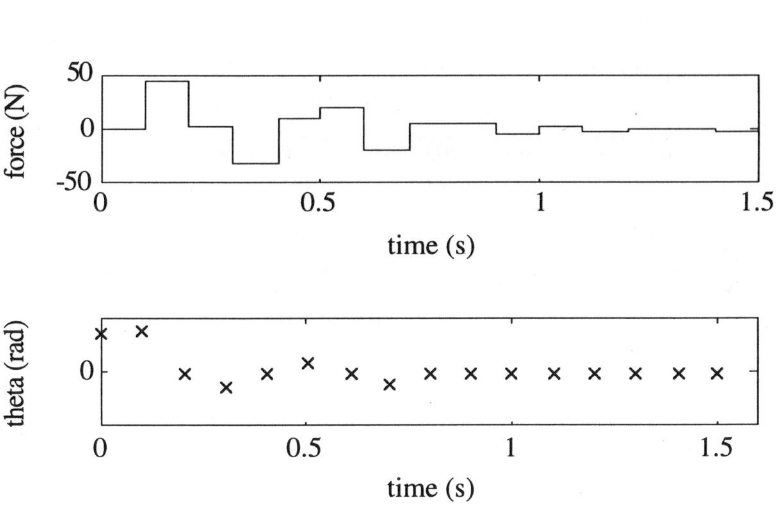

Fig. 10 Simulation for initial theta = 0.3.

11.5.2 CONTROL SYSTEM BEHAVIOR INCLUDING DISTURBANCES

In the previous section the control system was successfully implemented but no disturbances such as noise or faulty state measurements were added. Two different disturbance locations were investigated. Noise was be added after the plant, which represents a faulty measurement or sensor noise and the disturbance was to the controller output (actuator noise). The second case poses a larger problem for the controller. The disturbances added to the true value are random and the signal to noise ratio is 10. In the first case, random noise was added to the value of Θ, representing sensor noise. Afterwards, noise was added to the control for F to simulate actuator noise. In Figures 11 and 12 the behavior of the control system with the added disturbances are plotted. It can be seen that stability was achieved and that the added disturbances cause no significant degradation in performance

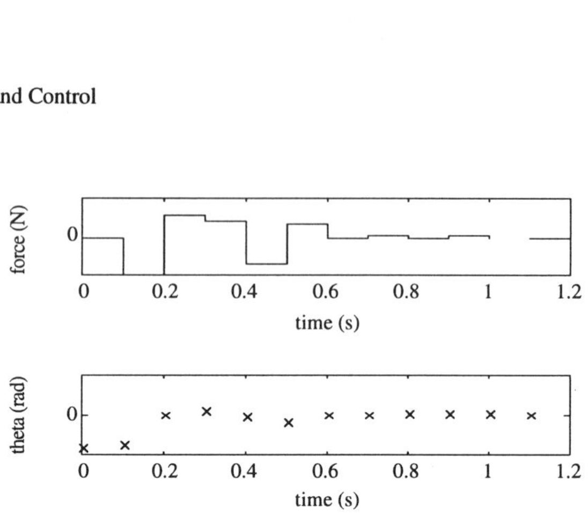

Fig. 11 Simulation with sensor noise.

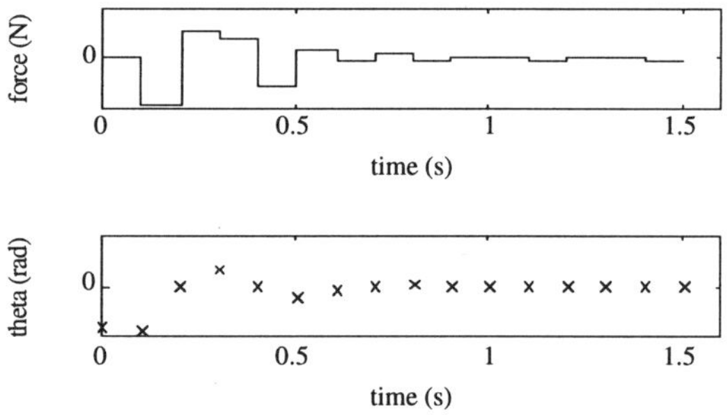

Fig. 12 Simulation with process noise.

11.5.3 CONTROL SYSTEM BEHAVIOR WITH THE TRUE PLANT

After achieving the goal of controlling the neural network plant and the disturbance rejection test, it is in addition necessary to demonstrate the controller performance with the real plant. This will also give an indication of the robustness of the controller to plant mismodelling since the neural network identified model is not perfect. For this reason the neural network controller output force for the present states Θ and $\dot{\Theta}$ is used in the simulation equation (Equation 12) and the new values for Θ and $\dot{\Theta}$ are calculated.

The neural network controller performs well and stability for the closed loop system is achieved. This result is very important since it justifies the approach and allows the controller design without using the true plant. Figure 13 shows the results achieved for the control of the true plant.

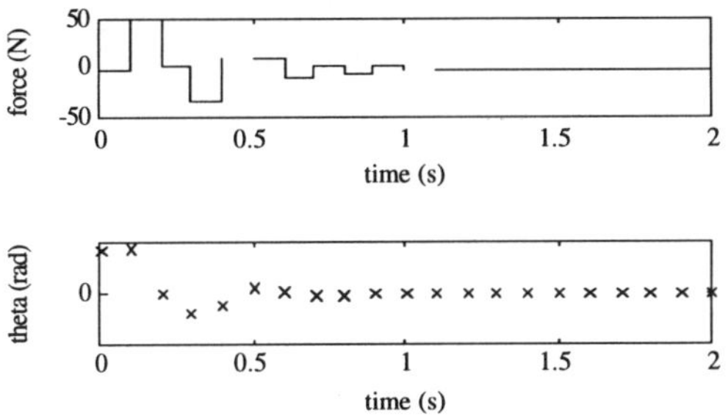

Fig. 13 Simulation for the true plant with initial theta = 0.3.

It is also important to note that the choice of enhancements was generic and that there was no enhancement that exactly matched the plant model. This is further confirmed by the fact that there was no enhancement weight determined by the identification which dominated the other weights. This behavior is critical since, as mentioned previously, the choice of enhancements cannot be determined based on the unknown plant model. The robustness to the choice of enhancements is thus revealed as well. This also shows s similarity between our approach and a nonlinear adaptive control approach since the weights on the enhancements can be considered parameter vector and the enhancement functions can be considered a nonlinear regressor vector. An important difference is that we rely on the functional link network to adjust the weights, rather than an explicit parameter update algorithm.

11.6 CONCLUSION

We demonstrated an effective method to use neural network structures in the controller design for linear or nonlinear plants. The approach is derived from a number of traditional control techniques. Two different neural networks were used to design a feedback controller for dynamic systems. The first serves as a system identification to identify the plant behavior. This plant model was then extensively used for simulations and as a consultant for the controller. The second neural network was the controller. After defining an objective function for the controller, appropriate data was collected and then used to train the controller network. The two-state neural network architecture may be used when the dynamic equations of the plant are not known. It is necessary though, to be able to observe, i.e. measure all plant states required for control. It can also be used to design multi-input multi-output neural network controllers. This neural network based control technique was tested for an inverted pendulum plant.

The results obtained for the control of the inverted pendulum show successful control system performance. The system identification is very important, since it models the true system behavior and predicts the output states for given inputs. Using different patterns with changing initial states, training of the neural network was achieved easily. Accurate modeling of the true plant allowed us to use the system identification neural network as a good approximation for the true plant. The controller neural network was trained using only a small number of patterns and the objective of balancing the pole to zero angle from various positions was achieved without difficulty.

Disturbances were added separately to the plant outputs and the controller output. The first case represented noisy sensors to measure the plant states and the second case adds disturbances to the control signal. The neural network controller performed extremely well and pole balancing was achieved in both cases.

Finally the neural network based controller was used to control the true plant. This verified the robustness of the controller to any mismodeling in the system identification neural network. The inverted pendulum, a nonlinear plant with inherently unstable dynamics, was successfully controlled.

11.7 APPENDIX

The 37 enhancements used in the FLN identifier are

$\Theta, \Theta, F, \Theta*F, \Theta*F, \Theta*\Theta, \sin\Theta, \sin\Theta, \cos\Theta, \cos\Theta, \Theta*\Theta*F, F*\sin\Theta, F*\sin\Theta, F*\cos\Theta,$

$F*\cos\dot{\Theta}, \Theta^2, \Theta^2*\dot{\Theta}, \dot{\Theta}*\sin\Theta, \dot{\Theta}*\sin\dot{\Theta}, \dot{\Theta}*\cos\Theta, \dot{\Theta}*\cos\dot{\Theta}, \dot{\Theta}^2*\Theta, \dot{\Theta}^2, \dot{\Theta}*\sin\Theta, \dot{\Theta}*\sin\Theta,$

$\Theta*\cos\Theta, \dot{\Theta}*\cos\dot{\Theta}, \Theta*\dot{\Theta}*\sin\Theta, \Theta*\dot{\Theta}*\sin\Theta, \Theta*\dot{\Theta}*\cos\Theta, \Theta*\Theta*\cos\Theta, \sin\Theta*\sin\dot{\Theta},$

$\sin\Theta*\cos\Theta, \sin\Theta*\cos\dot{\Theta}, \sin\dot{\Theta}*\cos\Theta, \sin\Theta*\cos\Theta, \cos\Theta*\cos\dot{\Theta}$

The 11 enhancements used in the FLN controller are

$\Theta, \dot{\Theta}, \Theta*\dot{\Theta}, \sin\Theta, \sin\dot{\Theta}, \cos\Theta, \cos\dot{\Theta}, \Theta*\cos\Theta, \dot{\Theta}*\sin\Theta, \Theta*\cos\Theta, \sin\Theta*\cos\Theta$

11.8 REFERENCES

[1] Atkeson,C.G.,"Memory-Based Learning Control," in Proceedings of the 1991 American Control Conference, pages 2131-2136, 1991.

[2] Barto,A.,Sutton,R., and Anderson,C.,"Neuronlike elements that can solve difficult learning control problems," in Transactions on Systems, Man, Cybernetics, pages 835-846, IEEE, 1983.

[3] Chen,F.,"Back-Propagation Neural Networks for Nonlinear Self-Tuning Adaptive Control," IEEE Control Systems Magazine, 10(3):44-48, 1990.

[4] Chen,V.C.,Learning Control with Neural Nets.PhD thesis, Case Western Reserve University, Department of Electrical Engineering and Applied Physics, January, 1989.

[5] Franklin,G.F.,Powell,J.D., and Workman,M.L., Digital Control of Dynamic Systems. 2nd Edition, Addison-Wesley, Reading, Massachusetts, 1990.

[6] Haesloop,D. and Holt,B.R.,"Neural Network for Process Identification," Proceedings of the IEEE International Conference on Neural Networks, 3(1):429-434. 1990.

[7] Hopfield,J.,"Neurons with Graded Response have collective Computational Properties like those of Two-State Neurons," Proceedings of the National Academy

of Sciences, 81(1):3088-3092, 1984.

[8] Hornik,K.,"Multilayer feedforward networks are universal approximations," Neural Networks Magazine, 2(1):303-314, 1989.

[9] Isidori,A., Nonlinear Control Systems, Springer-Verlag, 1989.

[10] Iwata,T., Machida,K., and Toda, Y.,"Fuzzy Control Using Neural Network Techniques," Proceedings of the IEEE International Conference on Neural Networks, 3(1):364-370, 1990.

[11] Jones,R.D.,"Function approximation with Neural Networks," in Proceedings of the IEEE International Conference on Neural Networks, pages 649-666, 1990.

[12] Kandel,A.,Fuzzy Mathematical Techniques with Applications. Addison-Wesley, Reading, Massachusetts, 1986.

[13] Kochenburger,R.J.,"A Frequency Response Method for Analysis and Synthesis of Contactor Servomechanisms," Transactions of AIEE, 69(1):270-284, 1950.

[14] Michalsky,R., Carbonell,J., and Mitchell,T., Machine Learning: An Artificial Intelligence Approach, Tioga, Palo Alto, California, 1983.

[15] Michie,D. and Chambers,R., Machine Intelligence: An experiment in adaptive control. Olin and Boyd, Edinburgh, 1988.

[16] Narendra,K.,"Identification and Control of Dynamical Systems Using Neural Networks," IEEE Transactions on Neural Networks, 1(1):4-27,1990.

[17] Nguyen,D.H. and Widrow,B.,"Neural Networks for Self-Learning Control Systems," IEEE Control Systems Magazine, 10(3):18-23, 1990.

[18] Ogata,K., Discrete Time Control Systems, Prentice-Hall, Englewood Cliffs, NJ, 1987.

[19] Ogata,K., "Modern Control Engineering," IEEE Control Systems Magazine, 6(4):56-74, 1970.

[20] Pao,Y.H., Adaptive Pattern Recognition and Neural Networks. Addison-Wesley, Reading, MA, 1989.

[21] Pao,Y.H., Klaasen,M., and Chen, V.,"Characteristics of the Functional Link Net: A Higher Order Delta Rule Net," Proceedings of the IEEE International Conference on Neural Networks, 1(1):507-514, 1988.

[22] Patrikar,A. and Provence,J.,"Self-Organizing Controller for Dynamic Processes using Neural Networks," Proceedings of the IEEE International Conference on Neural Networks, 3(1):359-364, 1990.

[23] Rumelhart,D., Learning internal representations by error propagation. Parallel Distributed Processing: Explorations in the Microstructures of Cognition, MIT Press, Cambridge, MA, 1986. Foundations, Vol. 1.

[24] Sandon, P.A.,"A Local Interaction for Adaptive Networks," Proceedings of the IEEE International Conference on Neural Networks, 1(1):317-324, 1988.

[25] Scott,R. and Collins, D., "Neural Network Adaptive Controllers," Proceedings of the IEEE International Conference on Neural Networks, 3(1):381-386, 1990.

[26] Shoureshi, R., "Learning and Decision Making for Intelligent Control Systems," IEEE Control Systems Magazine, 11(1):34-37, 1991.

[27] Vidyasagar, M., Nonlinear Systems Analysis. Prentice-Hall, Englewood Cliffs, NJ 1978.

[28] Wang,S. and Yeh, H., "Self-Adaptive Neural Architecture for Control Applications," Proceedings of the IEEE International Conference on Neural Networks, 3(1):309-314, 1990.

[29] Zwingelstein,G.M., Learning Neural-Net representation of nonlinear dynamic processes with the Functional Link Net. Master's thesis, Case Western Reserve University, Department of Electrical Engineering and Applied Physics, 1990.

Chapter 12

RAMSEY NUMBERS

KAZUHIRO TSUCHIYA
Fuji Electric, Japan
YOSHIYASU TAKEFUJI
Keio University, Japan

This paper describes a neural network parallel algorithm for finding Ramsey numbers in two-coloring. The Ramsey number, R(r,b) is given by the smallest value of N; R(r,b)=N where all edges of a complete graph of (N-1) vertices are colored either red or blue, and red-colored and blue-colored subgraphs should not form any complete subgraphs of r vertices and that of b vertices respectively. A superimposed red-colored and blue-colored subgraph is called a Ramsey graph. Our parallel algorithm based on the hysteresis McCulloch-Pitts neural network found five Ramsey numbers successfully. The neural network requires n(n-1) neurons for a Ramsey number of an n-vertex complete graph. The simulation result of the proposed algorithm is so promising that we may have a chance to find unknown Ramsey numbers.

12.1 INTRODUCTION

In 1928, the mathematician Frank P. Ramsey introduced "Ramsey Theory". Since then many scientists have widely studied the Ramsey theory [1]. To find the Ramsey numbers is one of the toughest problems in combinatorics. In spite of intensive study on the Ramsey numbers, only seven Ramsey numbers, R(3,3), R(3,4), R(3,5), R(3,6), R(3,7), R(3,9) and R(4,4) have been found [2][3]. For finding the Ramsey number, few algorithms have been reported [3] and no parallel algorithms have been given within our knowledge.

Hopfield and Tank first introduced a neural network model for solving combinatorial optimization problems [4]. In the artificial neural network model, a large number of neurons are used as processing elements, which perform the simplified biological neurons. They used the sigmoidal function as the neuron model. To enhance the convergence speed, Takefuji et al. have proposed the hysteresis McCulloch-Pitts neural network [5]. The input/output function of the hysteresis McCulloch-Pitts neuron model is given by:

$$\begin{aligned} V_{ijk} &= 1 \text{ if } U_{ijk} > \text{UTP (Upper Trip Point)} \\ &= 0 \text{ if } U_{ijk} < \text{LTP (Lower Trip Point)} \\ &\text{unchanged otherwise, for } i<j \text{ and LTP} < \text{UTP} \end{aligned} \tag{1}$$

where V_{ijk} and U_{ijk} are the output and input of the ijk-th processing element, respectively. The ijk-th processing element represents the state of an edge between the i-th vertex and the j-th vertex with k-th color. In this paper we have only two colors: k=1 or 2.

12.2 RAMSEY NUMBER PROBLEMS

12.2.1 PROBLEM DESCRIPTION

The Ramsey number, R(r,b) is given by the smallest value of N; R(r,b)=N where all edges of a complete graph of (N-1) vertices are colored either red or blue, and red-colored and blue-colored subgraphs should not form any complete subgraphs of r vertices and that of b vertices respectively. A superimposed red-colored and blue-colored subgraph is called a Ramsey graph.

According to the latest results for the Ramsey numbers [2][3][6], only seven Ramsey numbers, R(3,3), R(3,4), R(3,5), R(3,6), R(3,7), R(3,9) and R(4,4) have been found and fixed, and the others are still unknown.

12.2.2 SYSTEM REPRESENTATION AND ALGORITHM

We first define that V_{ijk} and U_{ijk} are the output and input of the ijk-th processing element, respectively and the ijk-th processing element represents the state of an edge between the i-th vertex and the j-th vertex with k-th color. $V_{ij1}=1$ means the ij-th edge is red-colored and $V_{ij2}=1$ blue-colored.

Fig. 1(a) shows a system representation for finding a Ramsey graph of the 5-vertex complete graph where each square represents one processing element. Two processing elements are used to color an edge between the i-th vertex and the j-th vertex. One is for the red edge and the other for blue. Generally $n \times (n-1)$ processing elements are necessary because $\frac{n(n-1)}{2}$ edges must be embedded in an n-vertex complete graph. In Fig. 1(a), twenty (=5(5-1)) processing elements are used.

Fig. 1(b) shows the solution state of processing elements for Ramsey number, (R(3,3)-1) which is 5. The black square and the white square show $V_{ijk} = 1$ and $V_{ijk} = 0$ respectively. The 13-rd edge, the 14-th edge, the 23-rd edge, 25-th edge, and the 45-th edge are red-colored (k=1), and the 12-nd edge, 15-th edge, the 24-th edge, the 34-th edge, and the 35-th edge are blue (k=2).

The state of the system in Fig. 1(b) represents the subgraphs in Fig. 1(c). The superimposed two-subgraph becomes the Ramsey graph of the 5-vertex complete graph.

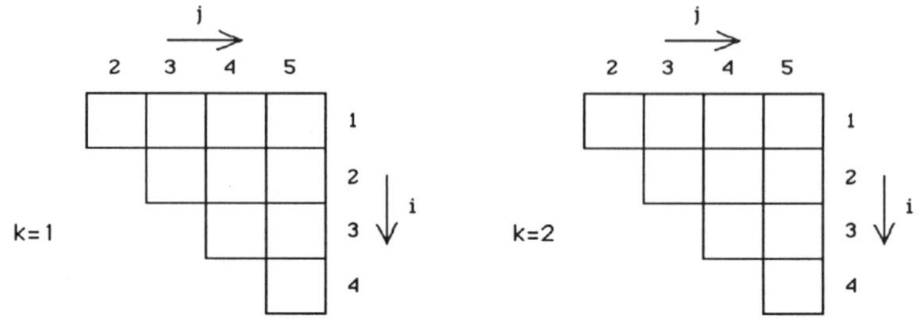

(a) Twenty Processing Elements for Finding a Ramsey Graph of Five Vertices.

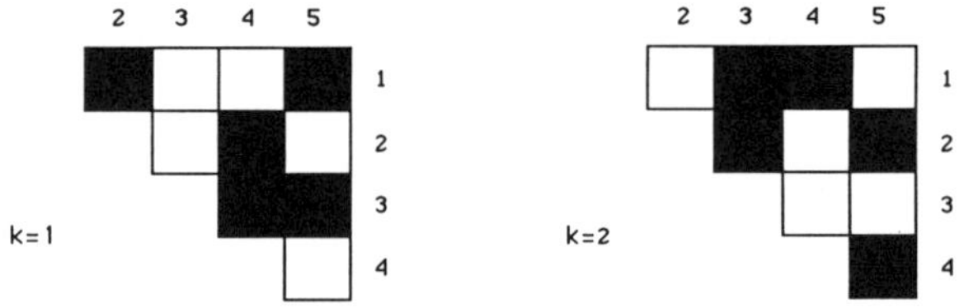

(b) The Solution State for (R(3,3)-1).

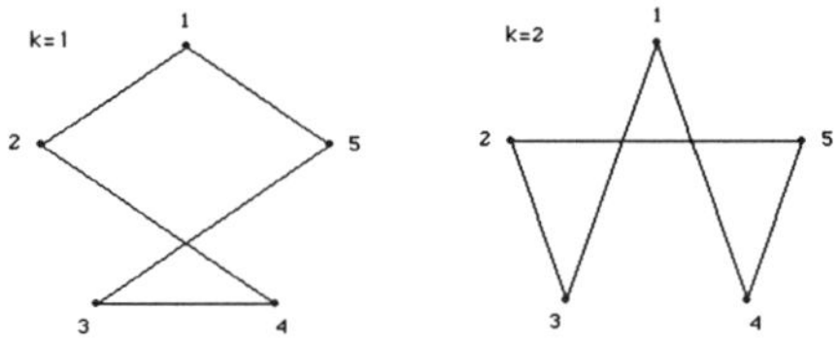

(c) The Corresponding Ramsey Graph

Figs. 1(a)–(c) System Representation for finding Ramsey Graphs.

The hysteresis McCulloch-Pitts neuron model is used to give the value of V_{ijk}. The change of the U_{ijk} is given by the following motion equation.

$$\frac{dU_{ijk}}{dt} = - \frac{\partial E(V_{121}, V_{122}, \ldots, V_{n-1n1}, V_{n-1n2})}{\partial V_{ijk}} \tag{3}$$

The energy function of the state, E is basically composed of two parts : color constraint and edge violation.

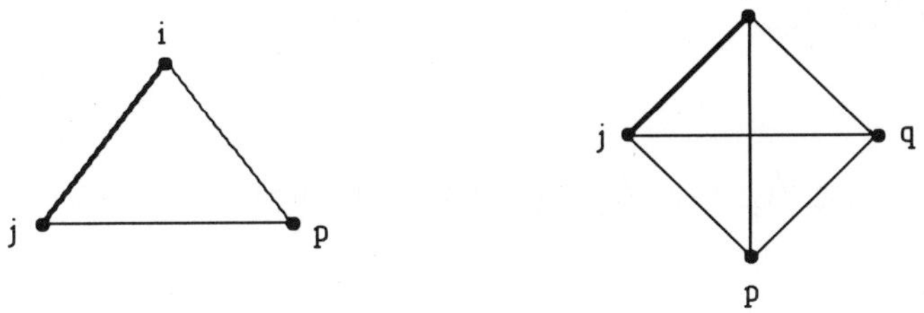

$V_{ijk} = V_{ipk} = V_{pjk}$

(a) Three Vertices

$V_{ijk} = V_{ipk} = V_{iqk} = V_{pjk} = V_{qjk} = V_{pqk}$

(b) Four Vertices

Fig. 2 Violation Condtions for the ijk-th edge.

The color constraint term for the ij-th edge is given by the following equation:

$$L[i,j,k] = \frac{1}{2}\left(\sum_{k=1}^{2} V_{ijk} - 1\right)^2 \tag{4}$$

This term is to force one and only one neuron to fire in the ij-th edge. In other words, the ij-th edge must be colored red or blue. If both two or no neuron is fired in the ij-th edge, then it becomes 1, 0 otherwise.

The edge violation conditions for the ijk-th edge are shown in Fig. 2. If the assignment of the ijk-th edge to the red (blue) subgraph creates complete subgraph of r (b) vertices, the ijk-th edge must not be embedded in it. These violation terms R[i,j,1] for red edges and B[i,j,2] for blue are given by:

$$R[i,j,1] = \sum_{\substack{p=1 \\ p \neq i,j}}^{n} V_{ip1} V_{pj1} V_{ij1} \qquad \text{for } r=3 \tag{5}$$

$$R[i,j,1] = \sum_{\substack{p=1 \\ p \neq i,j}}^{n} \sum_{\substack{q=p+1 \\ q \neq i,j}}^{n} V_{ip1} V_{iq1} V_{pj1} V_{qj1} V_{pq1} V_{ij1} \qquad \text{for } r=4 \tag{6}$$

$$B[i,j,2]=\sum_{\substack{p=1\\p\neq i,j}}^{n} V_{ip2}V_{pj2}V_{ij2} \qquad \text{for } b=3 \tag{7}$$

$$B[i,j,2]=\sum_{\substack{p=1\\p\neq i,j}}^{n}\sum_{\substack{q=p+1\\q\neq i,j}}^{n} V_{ip2}V_{iq2}V_{pj2}V_{qj2}V_{pq2}V_{ij2} \qquad \text{for } b=4 \tag{8}$$

$$B[i,j,2]=\sum_{\substack{p=1\\p\neq i,j}}^{n}\sum_{\substack{q=p+1\\q\neq i,j}}^{n}\sum_{\substack{r=q+1\\r\neq i,j}}^{n} V_{ip2}V_{iq2}V_{ir2}V_{pj2}V_{qj2}V_{rj2}V_{pq2}V_{pr2}V_{qr2}V_{ij2}$$
$$\text{for } b=5 \tag{9}$$

$$B[i,j,2]=\sum_{\substack{p=1\\p\neq i,j}}^{n}\sum_{\substack{q=p+1\\q\neq i,j}}^{n}\sum_{\substack{r=q+1\\r\neq i,j}}^{n}\sum_{\substack{s=r+1\\s\neq i,j}}^{n} V_{ip2}V_{iq2}V_{ir2}V_{is2}V_{pj2}V_{qj2}V_{rj2}V_{sj2}$$
$$\times V_{pq2}V_{pr2}V_{ps2}V_{qr2}V_{qs2}V_{rs2}V_{ij2} \qquad \text{for } b=6 \tag{10}$$

These conditions are nonzero if the assignment of the ijk-th edge to the red (blue) subgraph creates a complete subgraph of r (b) vertices.

The energy functions of the ijk-th processing element for finding a Ramsey graph of the n-vertex complete graph R(r,b) are given by:

$$E_{ij1}=\alpha\cdot L[i,j,1]+\beta\cdot R[i,j,1]-\gamma\cdot B[i,j,1] \tag{11}$$
$$E_{ij2}=\alpha\cdot L[i,j,2]-\beta\cdot R[i,j,2]+\gamma\cdot B[i,j,2] \tag{12}$$

where α, β, and γ are constant coefficients.

These motion equations of the ijk-th processing element are given by the equation (3) as follows:

$$\frac{dU_{ij1}}{dt}=-\alpha\cdot L'[i,j,1]-\beta\cdot R'[i,j,1]+\gamma\cdot B'[i,j,1] \tag{11'}$$

$$\frac{dU_{ij2}}{dt} = \alpha \cdot L'[i,j,2] + \beta \cdot R'[i,j,2] - \gamma \cdot B'[i,j,2] \tag{12'}$$

where

$$L'[i,j,k] = \sum_{k=1}^{2} Vijk-1 \tag{4'}$$

$$R'[i,j,1] = \sum_{\substack{p=1 \\ p \neq i,j}}^{n} V_{ip1} V_{pj1} \qquad \text{for } r=3 \tag{5'}$$

$$R'[i,j,1] = \sum_{\substack{p=1 \\ p \neq i,j}}^{n} \sum_{\substack{q=p+1 \\ q \neq i,j}}^{n} V_{ip1} V_{iq1} V_{pj1} V_{qj1} V_{pq1} \qquad \text{for } r=4 \tag{6'}$$

$$B'[i,j,2] = \sum_{\substack{p=1 \\ p \neq i,j}}^{n} V_{ip2} V_{pj2} \qquad \text{for } b=3 \tag{7'}$$

$$B'[i,j,2] = \sum_{\substack{p=1 \\ p \neq i,j}}^{n} \sum_{\substack{q=p+1 \\ q \neq i,j}}^{n} V_{ip2} V_{iq2} V_{pj2} V_{qj2} V_{pq2} \qquad \text{for } b=4 \tag{8'}$$

$$B'[i,j,2] = \sum_{\substack{p=1 \\ p \neq i,j}}^{n} \sum_{\substack{q=p+1 \\ q \neq i,j}}^{n} \sum_{\substack{r=q+1 \\ r \neq i,j}}^{n} V_{ip2} V_{iq2} V_{ir2} V_{pj2} V_{qj2} V_{rj2} V_{pq2} V_{pr2} V_{qr2}$$

$$\text{for } b=5 \tag{9'}$$

$$B'[i,j,2] = \sum_{\substack{p=1 \\ p \neq i,j}}^{n} \sum_{\substack{q=p+1 \\ q \neq i,j}}^{n} \sum_{\substack{r=q+1 \\ r \neq i,j}}^{n} \sum_{\substack{s=r+1 \\ s \neq i,j}}^{n} V_{ip2} V_{iq2} V_{ir2} V_{is2} V_{pj2} V_{qj2} V_{rj2} V_{sj2}$$
$$\times V_{pq2} V_{pr2} V_{ps2} V_{qr2} V_{qs2} V_{rs2} \qquad \text{for } b=6 \tag{10'}$$

The first term on the right side of Eq. (11') encourages the sum of the output of the ijk-th processing element to be one so that the ijk-th edge is colored by either red or blue. The second term encourages the output of the ij1-th processing element to be zero if the ij1-th edge creates a complete subgraph of r vertices. The third term encourages the output of the ij1-th processing element to be nonzero if the ij2-th edge creates a complete subgraph of b vertices.

In addition, we add the following hill-climbing term to each motion equation:

$$\omega \cdot H[i,j,k]=\omega \cdot h(\sum_{k=1}^{2} V_{ijk}) \tag{13}$$

where h(x)=1 if x=0, 0 otherwise and ω is a constant coefficient. This term also plays an important role to increase the frequency of the state of the system to converge to the global minimum [3].

In this paper, the synchronous parallel system was simulated on a Macintosh Powerbook170, although the algorithm is executable in a parallel machine. The synchronous parallel system can be performed on maximally n(n-1) processors for finding a Ramsey graph of the n-vertex complete graph. The state of n(n-1) processing elements can be updated by using Eqs. (11'), (12'), and (13) with the first order Euler method.

12.3 RESULTS AND DISCUSSION

The following five problems were solved by the algorithm.

(1) Problem #1: finding the Ramsey graph of 5-vertex complete graph for R(3,3).
(2) Problem #2: finding the Ramsey graph of 8-vertex complete graph for R(3,4).
(3) Problem #3: finding the Ramsey graph of 13-vertex complete graph for R(3,5).
(4) Problem #4: finding the Ramsey graph of 17-vertex complete graph for R(3,6).
(3) Problem #5: finding the Ramsey graph of 17-vertex complete graph for R(4,4).

Figures 3, 4, 5, and 6 show the simulation results for the problem #2, #3, #4, and #5 respectively. The algorithm found several Ramsey graphs from the different initial values of U_{ijk}. Table 1 shows the average number of iteration steps and the frequency of the convergence to solutions within 700 iteration steps, where 100 simulation runs were performed for each one of five problems. In every simulation run, the different initial values were given. Fig. 7 shows the relationship between the frequency and the number of iteration steps to solutions in the problem #1 and #3. The simulation results showed the consistency of the algorithm. The primary goal of finding the Ramsey graphs using our parallel algorithm was successfully achieved in terms of the computation time and the solution quality.

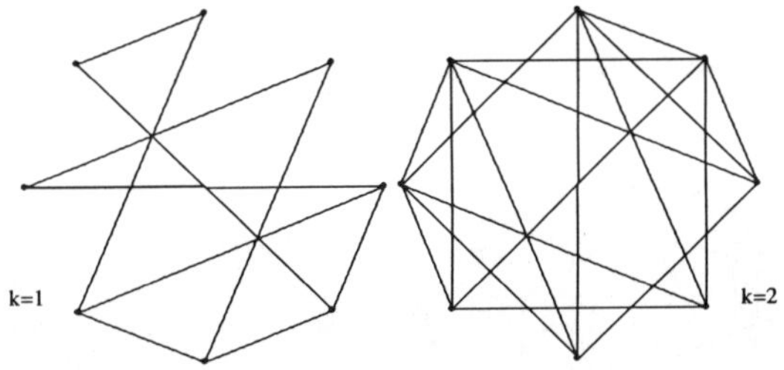

Fig. 3 A Ramsey Graph with 8 vertices for R(3,4).

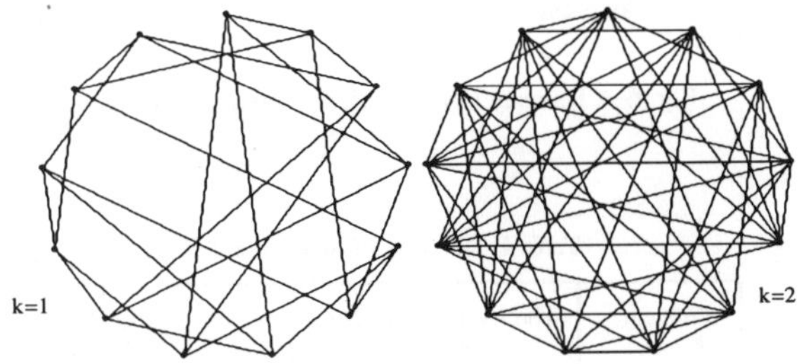

Fig. 4 A Ramsey Graph with 13 vertices for R(3,5).

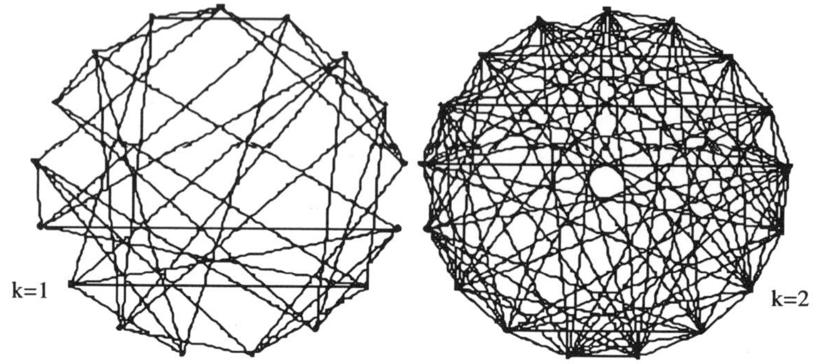

Fig. 5 A Ramsey Graph with 17 vertices for R(3,6).

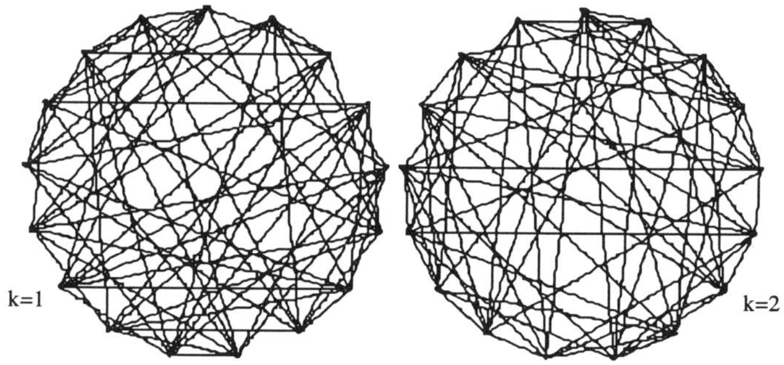

Fig. 6 A Ramsey Graph with 17 vertices for R(4,4).

Table 1 Summary of Simulation Results

Problem #	The Average Number of Iteration Steps	Convergence Frequency to Solutions
1. R(3,3)	28.7	100 %
2. R(3,4)	101.6	100 %
3. R(3,5)	248.9	85 %
4. R(3,6)	373.4	22 %
5. R(4,4)	361.0	7 %

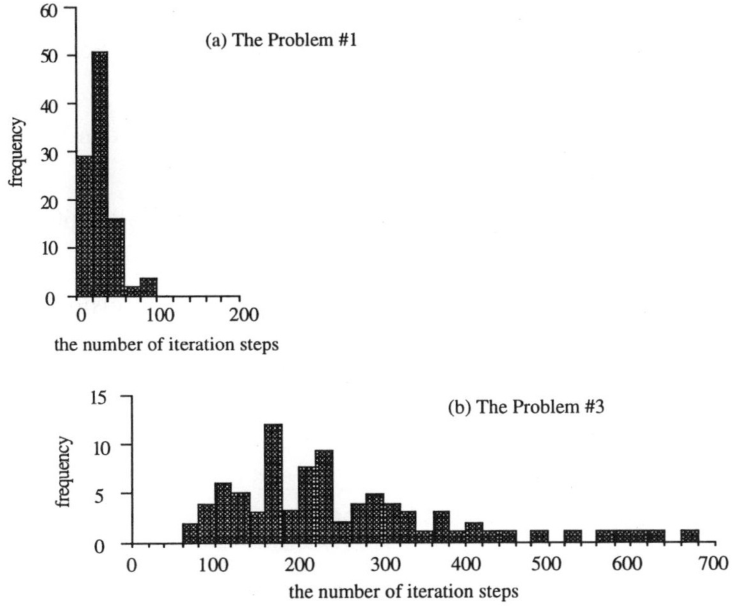

Fig. 7 Relationship between the frequency and the number of iteration steps.

12.4 CONCLUSION

The proposed parallel algorithm found Ramsey graphs of several complete graphs. The simulation results showed the consistency of our algorithm in terms of the computation time and the solution quality. The algorithm uses n(n-1) processing elements for finding a Ramsey graph of an n-vertex complete graph. The primary goal of finding the Ramsey graphs using the new motion equation was successfully achieved. We will extend the motion equations and the parallel algorithm to find new Ramsey numbers.

2.5 REFERENCES

[1] R. L. Graham, B. L. Rothschild, and J. H. Spencer, "Ramsey Theory," Scientific American, pp. 112-117, July 1990.

[2] R. L. Graham, B. L. Rothschild, and J. H. Spencer, Ramsey Theory, John Wiley & Sons, New York, 1980 and 1990.

[3] S. P. Radziszowski and D. L. Kreher, "Search Algorithm for Ramsey Graphs by Union of Group Orbits," J. Graph Theory, Vol. 12, 1, pp. 59-72, 1988.

[4] J. J. Hopfield and D. W. Tank, "Neural Computation of Decisions in Optimization Problems," Biological Cybernetics, 52, pp. 141-152, 1985.

[5] Y. Takefuji and K. C. Lee, "An Artificial Hysteresis Binary Neuron: A Model Suppressing the Oscillatory Behaviours of Neural Dynamics," Biological Cybernetics, 64, pp. 353-356, 1991.

[6] G. Polya, R.E. Tarjan, and D.R. Woods, Notes on Introductory Combinatorics, Birkhauser Boston, Inc. 1983.